U0156046

阅读
南京路

乔争月 著
Michelle Qiao

上海三联书店

Shanghai
Nanjing Road

Ricsha trip through Nanjing Road in 1925

For the traveler who wants to make a thorough tour of the town, or as thorough as his limited time will allow, there is no better way than to get in a ricsha and follow several of the most interesting thoroughfares, getting off from the vehicles at time and exploring some of the more obscure places.

Except from an article named "Ricsha Trips Through City Worth While; Good Routes, Nanking Road Principal Thoroughfare in Shanghai" on the *China Press* in 1925.

1925年乘黄包车漫游南京路

摘自 1925 年《大陆报》文章，题为《乘黄包车穿越城市值得一游，南京路——上海第一要道》

对于想深度游览上海或在有限时间内深入探索的旅行者，没有比乘坐黄包车沿几条最有趣的街道游览更好的方式了。可以随时下车去探索一些更冷门的地方。

西藏路
Thibet Road

浙江路
Chekiang Road

The following routes will be found to be among the most interesting for the tourist on the Empress of France. The itinerary starts at the corner of the Bund and Nanking Road and proceed is along the latter road.

Starting west we past first several curio shops on the left hand side, then the section of the foreign department stores is reached. In the section most of the foreigners' requirements in the matter of clothing and outfitting can be attended to. A number of excellent lace and embroidery shops are also to be found here. Foreign tea-shops and confectionery stores follow. This section extends for three long blocks.

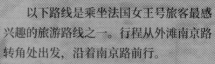

　　以下路线是乘坐法国女王号旅客最感兴趣的旅游路线之一。行程从外滩南京路转角处出发，沿着南京路前行。

　　从西端开始，我们先经过左边的几家古玩店，然后到达外国百货公司的区域。这一带的商店可以满足大多数外国人在服装方面的需求。这里还有许多出色的花边和刺绣商店。随后是西式茶室和糖果甜品店。这一区域延伸了三个街区。

Passing Honan Road notice on the right a very shabby-looking old building with a strange air of reserve and secrecy about it. No 49. This is one of the oldest foreign buildings in the Settlement. Many years ago when Shanghai was young and land was cheap, a number of the younger men of the Settlement formed themselves into a bowling club and erected their club-house here, "The Bowling Alley Club".

The town grew, the land soared in value, but the club remained and there once or twice a week some of the very oldtimers still come and seem to greatly enjoy the privilege of playing skitties on one of the most valuable pieces of land in the Settlement. The club is one of the most exclusive in Shanghai, 24 members only, and to remove one single cobweb from the dusty rafters of their house in the eyes of the members treason of the blackest.

　　过了河南路，请留意右边一幢相当破旧的老建筑，带有一种奇怪的保守的秘密气氛。门牌是 49 号。这是租界里最古老的外国建筑之一。许多年前，在租界早期土地便宜时，许多在租界生活的年轻人组建了一个保龄球俱乐部，并在这里建立了自己的俱乐部，即"保龄球场"。

　　上海城市发展后土地价值飞涨，但这个俱乐部仍然存在，老朋友们每周会在这里举行一两次活动，他们似乎很享受在租界最贵的土地上玩乐的特权。这也是上海最独特的俱乐部之一，仅有 24 名会员。如果要从屋里满是灰尘的橡子上去除一只蜘蛛网，在这些会员眼中就是最严重的背叛行为。

Here also, next door, the visitor will receive an idea of what the Ma-loo of ten or twenty years ago looked like. The gorgeous gilded and carved shop front with its sharply up-curving roofs and balconies, and intricately fretted jalousies, and general Chuchinchow atmosphere, is one of the last survivors of this delightfully attractive type of building in the street. Modern buildings with an abundance of stucco in Chinese designs, aptly dubbed "compradoresque", are unfortunately usurping command of the street.

We are now in the district of Chinese shops. Notice the food shops with their displays of pressed or varnished ducks, and other weird and fearsome looking confections, next see the big silk and silver hongs. These should be visited. Two of the best known Chinese silk shops in the Settlement, Laou Kiu Chwang and Leou Kiu Luen will be found on opposite sides of the street within a short distance of each other in big modern buildings.

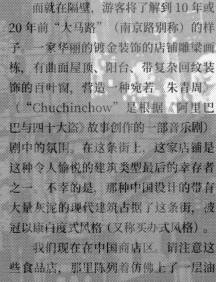

而就在隔壁，游客将了解到 10 年或 20 年前"大马路"（南京路别称）的样子。一家华丽的镀金装饰的店铺雕梁画栋，有曲面屋顶、阳台、带复杂回纹装饰的百叶窗，营造一种宛若《朱青周》（"Chuchinchow"是根据《阿里巴巴与四十大盗》故事创作的一部音乐剧）剧中的氛围。在这条街上，这家店铺是这种令人愉悦的建筑类型最后的幸存者之一。不幸的是，那种中国设计的带有大量灰泥的现代建筑占据了这条街，被冠以康白度式风格（又称买办式风格）。

我们现在在中国商店区。请注意这些食品店，那里陈列着仿佛上了一层油的鸭子，还有其他看起来很奇怪的令人害怕的点心。接下来可以看到经营丝绸和银制品的商行。这些店值得一逛。街对面就是租界里最著名的两家中国丝绸商店——Laou Kiu Chwang 和 Leou Kiu Luen。两家店位于大型现代建筑中，彼此相距不远。

Here on the right hand side is a "joss-goods" shop, incense sticks, joss-paper, imitation sycee, silver dollars, clothing, furniture, etc., for the use of the department "devotional articles" of all sorts. There is a reason for this shop's situation. Right next door we turn down a rather dark entry and drop a couple of thousand years or so in a minute. Here in the middle of the bustle and hurry of the busiest street in the Orient we are suddenly in the semi-dark, incense-laden atmosphere of a temple, Hung Miao. A prosperous, well-patronized Chinese merchant hurries in for a moment to arrange matters with his favorite diety--at so much per matter to the bland-faced priest. Notice that fat old compradore, with the twinkle in his eye and the large cigar just emerging. He has been in to "chin-chin joss" to make sure that the deal he has in mind with a foreign hong will be all his way. The chief gods represented here are Kwanyin, the goddess of Mercy, and Midoh and Waydoh, the coming and present Buddhas, with many lesser luminaries.

　　到这里，右侧是一家香烛店，销售燃香、香纸、纸元宝、纸银元和纸衣服家具等，都是用于虔诚供奉使用的物品。这家店经营这些商品是有原因的。就在店的隔壁，我们转进一个比较暗的入口，仿佛在一分钟内穿越了数千年。在东方最繁忙街道那车水马龙的中心，我们突然陷入了一个半黑暗的、香气缭绕的庙宇空间，就是红庙。

　　一位看起来生意很好的中国商人匆忙进来按照自己喜欢的方式来办事，而这对面无表情的法师来说也是如此。请注意这位肥胖的老买办，他的眼睛闪烁着，拿出一支大雪茄。他一直在拜佛，以确保他计划与外国洋行达成的交易可以心满意足。这里供朝拜的主要佛像是保佑慈悲的观世音菩萨，以及象征来世和现世的弥勒佛和韦陀。

Passing the Chekiang Road, a tremendously busy cross-town thoroughfare, we are between the splendid department stores of the Wing On Company and the Sincere Company. Both these are owned and operated entirely by Chinese.

After passing this part of Nanking Road the traffic becomes less congested. On the left side of the road, the next building of interest is the Town Hall, which houses the market and the public library.

过了交通非常繁忙的浙江路，我们就到了两家很出色的百货公司之间。这两家公司——永安百货和先施公司都是中国人开办经营的。

过了南京路这一段后，交通变得不那么拥挤了。在这条路的左侧，下一个有趣的建筑是市政厅，里面设有市场和公共图书馆。

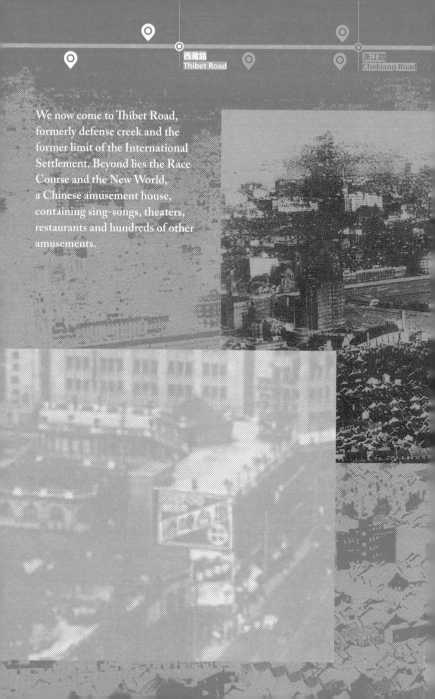

西藏路
Thibet Road

浙江路
Chekiang Road

We now come to Thibet Road,
formerly defense creek and the
former limit of the International
Settlement. Beyond lies the Race
Course and the New World,
a Chinese amusement house,
containing sing-songs, theaters,
restaurants and hundreds of other
amusements.

河南路
Honan Road

外滩南京路转角
The corner of the Bund and
Nanking Road

　　现在，我们来到西藏路，这条路原来是一条河浜，也是公共租界的界河。过了这条路，就是上海跑马场和新世界。后者是一家中国娱乐场所，提供有歌唱表演、剧院、餐厅和其他数百种娱乐活动。

阅读南京路

著者：乔争月

摄影：张雪飞 邵律 乔争月

插图：廖 方

鸣谢：

郑时龄 熊月之 邢建榕 章 明
唐玉恩 李乐曾 常 青 钱宗灏
王 健 童 明 淳 子 陈中伟
华霞虹 张姚俊 沈晓明 陈 贤

上海市档案馆
上海市政府新闻办公室
上海市图书馆
上海市徐家汇藏书楼
上海市文化和旅游局
上海市历史博物馆
上海市体育总会
上海音乐厅
上海市建筑学会
上海信托
上海市黄浦区人民政府档案馆
上海市黄浦区南京东路街道
上海市黄浦区外滩街道
上海市黄浦区南京路步行街管理办公室
上海市黄浦区置地（集团）有限公司
长征医院
同济大学校史馆
格致中学
上海市商贸旅游学校
百联集团
上海外滩投资发展有限公司
上海章明建筑设计事务所
匈牙利驻沪总领馆
上海日报
墨辰文化传媒有限公司
上海缪诗肖像摄影

Shanghai Nanjing Road

Author:Michelle Qiao

Photographer:Zhang Xuefei, Shao Lv, Michelle Qiao

Illustration:Liao Fang

Acknowledgements:

Zheng Shiling, Xiong Yuezhi, Xing Jianrong, Zhang Ming, Tang Yuen, Li Leceng, Chang Qing, Qian Zonghao, Wang Jian, Tong Ming, Chunzi, Chen Zhongwei, Hua Xiahong, Zhang YaoJun, Shen Xiaoming, Chen Xian

Shanghai Municipal Archives Bureau
Information Office of Shanghai Municipality
Shanghai Library
The Xujiahui Library
Shanghai Culture and Tourism Bureau
Shanghai History Museum
Shanghai Sports Club
Shanghai Concert Hall
The Architectural Society of Shanghai, China
Shanghai Trust
Shanghai Huangpu District Archives Bureau
The East Nanjing Road Sub-district Office of Huangpu District Shanghai Municipality
The Bund Sub-district Office of Huangpu District Shanghai Municipality
The Nanjing Pedestrian Road Management Office of Huangpu District Shanghai Municipality
Shanghai Huangpu District Property Co. Ltd.
Changzheng Hospital
History Museum of Tongji University
Gezhi High School
Shanghai Business & Tourism School
Bailian Group
Shanghai Bund Investment (Group) Co., Ltd.
Shanghai Zhangming Architectural Design Firm
Consulate General of Hungary in Shanghai
Shanghai Daily
Mochen Culture Media Co., Ltd.
MUSEE FOTO

南京西路新之商贸大楼鸟瞰 俞 二〇〇八年九月

目录

Contents

阅读《阅读南京路》

专注于上海建筑和城市空间的考证，重塑上海城市的历史记忆，在完成《武康路建筑地图》《外滩·上海梦》之后，乔争月女士的《阅读南京路》又将与读者见面。乔女士不仅是新闻媒体报导上海和上海建筑的专栏作家，也是研究上海近代建筑的专家，发表了许多文章和专著，对认知上海的历史建筑有很大的贡献。她从深藏在汗牛充栋的各种历史档案和外文文献中寻访，探究上海的建筑及其背后的历史和事件，并进行实地考察，绘声绘色地加以描述，带领我们认知昔日的上海和今天的上海。这一次，她把视线集中到号称中华第一街的南京路。

《阅读南京路》开宗名义就介绍了《大陆报》的一篇"1925年乘黄包车漫游南京路"的报导，领着我们穿越到当年的南京路。早在19世纪初，上海已有"江海之通津，东南之都会"之誉，当年上海的发展主要集中在老城厢内，已由早期的蕞尔小邑一跃而为东南名邑。上海控江踞海的地理位置，优越的港口条件，特殊的历史因素引起了海内外的注意。1843年上海开埠，苏格兰植物学家和旅行家富钧（Robert Fortune, 1812–1880）曾经在当年年底访问上海，预言上海"会成为一个更加重要的地方……它在许多地方优于南方的那些竞争对手。"

次年初，上海已有11家洋行开业，在上海外滩沿江一带划地盖楼。上海开埠两年之际，清政府公布了《上海租地章程》（Land Regulation），成为上海设立外国租界的法律依据，1846年建立的英租界东面以黄浦江为界，西面以界路（今河南中路）为界，这一时期的上海除华界之外，以租界的发展作为城市发展的核心。

南京路开始只是一条乡间小路，称作花园弄，到1865年，由26条道路组成的道路网已在英租界形成，花园弄改名为南京路，俗称大马路，开始使用煤气灯照明。当年沿街的大部分房屋仍为木构建筑，有着繁复细致的木雕栏杆、檐口和垂花装饰，底层作为商铺，二层是住家，很少见到三层以上的楼房。近百年来的发展使这一带成为上海以至全国最繁华的商业大街，商业更趋繁荣，大量消费品从国外涌进南京路，两侧的商店逐渐向西延伸。到 20 世纪 20 年代，南京路已形成繁华的商业中心，聚集了 30 余种行业约 200 家专业商店，使南京路享有"小巴黎"的美誉。四大公司的开设更是大都市商业现代化的标志，南京路也因此成为上海，乃至远东的商业中心，成为全世界最有趣的街道之一。

南京路在 1852 年成为中国的第一条柏油马路，1874 年的南京路上有了人力车，与传统的轿子、马车并驾齐驱。1901 年南京路上出现了上海有史以来的第一辆汽车，1908 年，第一辆有轨电车出现在南京路上，从外滩到西藏路一段全部铺上印度铁藜木，成为远东最漂亮的一条街

道。1914 年，无轨电车通车，1922 年，上海第一条公共汽车线路从南京西路站始发，通至兆丰公园（今中山公园）。这条从外滩直抵静安寺全长 5465 米的南京路两旁汇集了汇中饭店、沙逊大厦、国际饭店以及大华饭店等名楼，有着张园、哈同花园等园林宅邸，又有着新世界游乐场、四大公司的屋顶乐园、仙乐剧场、新新舞台、卡尔登大戏院、美琪大戏院、百乐门舞厅等娱乐场所，见证着南京路的百年繁华。

聚焦黄浦区段南京路和人民广场的《阅读南京路》告诉了我们 33 座有故事的建筑的前世今生，其中南京东路 16 座，南京西路 5 座，南京路周边地块的 12 座建筑，其中有百货公司、酒店、办公楼、银行、娱乐建筑、教会建筑、学校、公寓等，既有新建筑，也有老建筑。作者也以第一手文献资料生动地描写了建筑师和当年与这些建筑相交集的风云人物的故事，从独特的时空交错的视角研究上海和上海建筑，行文雅俗可读共赏，中英文并列，旁征博引，娓娓讲述，成为作者特有

的文风。

近年来，中外社会各界对历史档案、建筑、文物、地图和文献等进行了广泛的考辨、挖掘和研究，这些研究以涓涓细流汇成蔚为壮观的上海研究的大江河海，就像一幅巨大的拼图，社会各界都努力参与，贡献自己的力量。尽管这幅拼图还有许多空白，但这幅拼图的轮廓已经越来越清晰，《阅读南京路》也是这幅拼图的一个组成部分，我们翘首以待作者下一块拼图的到来。

乔争月女士的著作不是单纯的历史故事，更是展望未来的启示，为上海这座城市洒落温润的月光。

熊月之

2020 年 6 月 11 日

Reading "Shanghai Nanjing Road"

Recounting Shanghai's historical memories through continuous efforts on the research of Shanghai's architecture and urban spaces, the author Ms. Qiao Zhengyue (Michelle Qiao) is an excellent historical architecture columnist of Shanghai Daily and an expert of Shanghai's modern architecture. With her articles and books, she has done a great contribution to promote the understanding of Shanghai historical buildings. To find out the history and events behind these buildings, she has researched a large amount of Chinese historical archives and English documents, and done a lot of field visits. With her vivid descriptions, she helped us to get a better picture of Shanghai's past and present. After publishing books like "Shanghai Wukang Road" and "Shanghai Bund, Shanghai Dream," her new book "Shanghai Nanjing Road" is now ready for the readers. This time her focus is on the Nanjing Road, which has been widely known as the No. 1 Street of China.

The book "Shanghai Nanjing Road" starts from a report in 1925 from The China Press about a rickshaw trip through Nanjing Road, giving us a kaleidoscope view of the street at that time. In the early 19th century, Shanghai had already been known as "a good port at the cross of the East China Sea and the Yangtze River, a nice county in the southeast of China". At that time, though Shanghai had become well-known county in the southeast China, its main development was within the old city of the county. However Shanghai's ideal geographic location for a potential great port city under the then special historical background had already attracted attentions both from home and abroad. After Shanghai officially opened its port to the foreigners in 1843, the Scottish botanist/traveler Robert Fortune (1812-1880) visited the city at the end of the year. He predicted that Shanghai "will become a much more important place… it is in many places better than those of the South."

At the beginning of the following year, a rainbow of 11 foreign companies opened business branches in Shanghai and built buildings along the Shanghai Bund. Two years after Shanghai became an open port, the Qing government promulgated the "Shanghai Land Regulations" which became the legal basis for Shanghai to establish foreign settlements. The British settlement established in 1846 was bounded by the Huangpu River on the east and today's Henan Road M. on the west. During this period, besides the old city, the settlement area gradually became the new focus of the city's urban development.

Nanjing Road was initially a small country road called the Garden Lane. By 1865, a transport network consisting of 26 roads had been built in the British settlement. The renamed Nanjing Road, also commonly known as the "Ta Maloo" was part of the network and began to use gas lamps for lighting. At that time, most houses along the road were still wooden structures with complicated and refined wood carving railings, cornices and pendant decorations.

The ground floor was used as shops while the second floor used for living purpose. Buildings above three floors were rarely seen. As the business became more and more prosperous, more and more foreign goods showed up onto Nanjing Road, and the shops on both sides gradually extended westward. By the 1920s, with around 200 professional shops of more than 30 industries, Nanjing Road had become a prosperous commercial center with a nice reputation as the "Little Paris". The opening of the four big Chinese department stores on the road was also a milestone of the modernization of the metropolitan business. Therefore after almost a century's development, Nanjing Road became the most prosperous area in Shanghai, the commercial center of China and even the Far East Asia, and one of the most interesting streets in the world.

Nanjing Road became China's first asphalt road in 1852. Rickshaws showed up on Nanjing Road in 1874, along with traditional sedans and horse-drawn carriages at the same time. In 1901, the road witnessed the first car in Shanghai. In 1908, Shanghai's first tram was introduced into the road traffic. The section from the Bund to today's Tibet Road was all paved with Indian lignum vitae wood and became the most beautiful street in the Far East. In 1914, the trolley bus was put into use on this road. In 1922, the first bus line in Shanghai started operation from Nanjing Road W. Station to Jessfield Park (today's Zhongshan Park). From the Bund directly to Jing'an Temple, the 5465-meter-long Nanjing Road was flanked by famous hotels such as the Palace Hotel, the Sassoon House, the Park Hotel and the Majestic Hotel, by nice gardens and residences such as the Zhangyuan Garden and Aili Garden, and by wonderful entertainment venues such as the Great World entertainment center, the rooftop amusement parks of the big four department stores, the Sun Sun Stage, the Ciros Plaza, the Carlton Theatre, the Majestic Theatre, and the Paramount Ballroom, all of which had witnessed the prosperity of Nanjing Road over a century's time.

The book "Shanghai Nanjing Road" introduces us the past and the present of 33 historical buildings on Nanjing Road within today's Huangpu District, including 16 on Nanjing Road E., 5 on Nanjing Road W. and 12 around this area. Among them are department stores, hotels, office buildings, banks, entertainment buildings, Christian buildings, schools, apartments and all are with latest updates. With a smart use of first-hand documents and a unique perspective of the combination of the past and present of Shanghai and its historical architecture, the author vividly tells the stories of the architects and the historical figures whose lives and work once intertwined with these buildings. The well-documented bilingual writing which appeals to both the academic and the public has also become her unique style.

In recent years, huge efforts from both home and abroad have been made on Shanghai themed study through extensive examination, excavation, and research work on historical archives, buildings, cultural relics, maps and documents. These studies from different fields have all trickled together into an ever-growing mainstream of Shanghai study. It's also like a huge jigsaw puzzle game with active participations and contributions from all sectors of the society. Although there are still many blanks in this puzzle, the outline of it has been clearer than ever. With this book, "Shanghai Nanjing Road" being a new piece of the puzzle, we do look forward to the author's next piece in the near future.

Ms. Qiao Zhengyue's work is not only about telling history stories, moreover it is an enlightenment to the future like a beam of soothing moonlight for the city of Shanghai.

Zheng Shiling
June 11, 2020

南京路的新时代

2018年4月一个有风的晚上，我应上海市政府外事办邀请，陪伴访沪的新加坡总理李显龙夫妇漫步外滩。

夜色正好，我们从黄浦公园出发，沿滨江大道边走边聊，来到陈毅广场。看到璀璨的南京路，我问李总理，"您听说过南京路吗？这是上海的'第五大道'。"他回答，"我听说过这条路，它很有名，有很多商店，可惜我没有在上面走过。"

南京路游人如织，出于安保工作考虑，李总理确实难有机会一逛。我们所在的滨江大道与南京路隔着马路，他为了看得更清楚爬上了花坛，凝望这条夜色迷人的著名马路，还拿出相机拍了不少照片。

我想，这就是南京路的魅力。

让我们穿越到1929年的上海，站在让李总理流连忘返的外滩南京路口。根据同年一张西人绘制的上海地价图，包括大部分外滩和南京东路的这个T字型路口，是全市地价最高的黄金地段。

1843年上海开埠后，英国总领事巴富尔和上海道台宫慕久共同圈定位于老城外一块江边滩地为英租界。短短几年内，这块布满棉田和坟地的泥滩就升起一座迷你的欧洲城市，泥滩慢慢变成了"金滩"。

南京路开始则是一条外侨散步遛马的乡间小路，被叫作"马路"，正式的名字是"花园弄"（Park Lane）。西人爱骑马，每年的跑马季是节日盛会，女士们三天的盛装不会重样。1850年，外侨在今南京东路河南路一带的花园建跑马总会，在花园两侧设抛球场，沿花园和抛球场筑跑马道，就是上海的第一个跑马场。

不久，外滩因商业发展向西扩展，1854年跑马总会出让土地，在今浙江路西藏路之间建第二跑马场，1864年又让地，到今天的西藏中路西侧建第三跑马场，就是现在的人民公园和人民广场。跑马总会在10年间两次西移，反映当时租界扩展之迅速。

1921年上海工部局编写的《上海史》透露，英国传教士麦华陀（Walter Henry Medhurst）建议租界路名应让中国老百姓易于理解，他们中很多人为躲避太平天国战乱从各地逃到租界。此后，"花园弄"以中国古代首都南京命名为"南京路"。而这条从外滩通往跑马场的道路，也从一条乡间小路发展为远东最著名的商业街。1928年，卜舫济（F. L. Hawks Pott）在《上海简史》（*A Short History of Shanghai*）一书中写道，"到了晚上，这些商店被成千上万的电灯照亮，使南京路东部成为一条白色的大道。最重要的是，这些商店满足了需求，挤满了购物者，其中许多是来自其他城市的游客。这既标志着中国现代化的逐步发展，也象征着上海的日益繁荣。1934年版的英文《上海指南》（*"All About Shanghai A Standard Guidebook"*）则提到，上海这条主要的零售商业街被一位著名美国作家命名为"全世界最有趣的七个街道之一"。

1945年时，政府将从跑马场通往静安寺的静安寺路命名为南京西路，南京路东段则称为南京东路。整条南京路绵延达5公里多，其"十里洋场"的美名成为近代上海大都市的象征和代名词。

不过，南京路不全是洋人街。这条街道的城市空间由华洋长期接触和竞争后形成，以河南路为界，靠外滩的一段比较"洋气"，近人民广场一段的中国元素更多。

外滩一行洋楼里，只有一个中国建筑师参与的作品——陆谦受与公和洋行设计的中国银行。而在南京路，中国第一代建筑师展露光芒。他们在南京路的设计实践中，巧妙结合现代建筑技术与传统中国美学，留下很多经典作品，如八仙桥青年会大楼和大新公司。值得一提的是，投资兴建南京路建筑的中国业主数量也远远多于外滩。

上海这座城市最迷人之处，就是19世纪中叶后，中西文化曾在这里相遇、碰撞、竞争并互相学习、吸收，最后水乳交融。这也是南京路最让人心动的魅力。

据说上海早期的外国侨民

认为外滩是一把弓，而南京路是一支箭，一路向西，射向近代上海城市发展的方向。我的《上海外滩建筑地图》推出一年后的 2017 年初，上海市黄浦区人民政府开始对南京路进行整体改造，加速了这条著名历史街道在新时代的转型。

为了吸引年轻消费者并恢复"昔日的辉煌"，南京路的几家百年老店，如惠罗公司、永安百货和由大新公司改造的第一百货等，陆续进行保护修缮和功能提升。犹太地产商投资兴建的梦幻拱廊——中央商场，被改造为时尚的"外滩·中央"项目，引入"乔丹"和"林肯爵士乐"等国际知名品牌；而装饰艺术风格的上海电力公司和被定义为后现代经典的华东电管大楼，则合二为一，翻建成低调奢华的艾迪逊酒店。在曾经是一座"丝绸宫殿"的老介福大楼，华为开出全球最大的旗舰店，面积达 5000 平方米，是目前最大的华为店的三倍以上。

而最让人惊喜的是上海跑马场俱乐部历经多年修缮，改造为上海历史博物馆对公众开放。大楼既保留马头形铸铁栏杆等跑马场时代的历史细节，又增加了现代博物馆的设施。博物馆里还陈列着很多与南京路历史有关的珍贵文物，如百年前南京路翻修时铺设的铁藜木砖，那是"红木马路"名声的来源。我很有幸，戴着一顶顶颜色各异的安全帽，见证了这些南京路建筑历史更新的过程。

2019 年岁末，南京路步行街在庆祝 20 周年生日庆典后，迎来又一次升级改造——从河南中路开始延伸至四川中路。工程还把马路路面抬高至人行道水平，同时铺装石材以统一南京路的外观风格。这些 21 世纪的城市更新项目，给这条建于 19 世纪的马路，赋予了新的形象和传奇。

沿着百年前赛马奔跑的方向，我从外滩出发，穿过城市更新中尘土飞扬的南京路，直到上海的城市之心——人民广场。这沿途的风景和故事超乎想象，甚至比外滩还要精彩。

乔争月

2020 年 6 月于上海武康路月亮书房

The New Era of Nanjing Road

On the night of April 11, 2018, I was invited to accompany the Singaporean Prime Minister Mr. Lee Hsien Loong and his wife for a walk along the Bund.

Setting off from the Monument to the People's Heroes, I introduced to them the history of the bund along the road until we reached Chen Yi Square and saw the illuminated Nanjing Road. I asked the prime minister, "have you heard of Nanjing Road? It was Shanghai's 'Fifth Avenue.' "Oh, I've heard of this famous road. It has many shops. But it's a pity that I have never walked on it," he said.

Nanjing Road was always thronged with tourists which was not a good choice of walk for the prime minister. For a closer look at Nanjing Road which was opposite the street from where we stood on the bund promenade, he stepped up onto flower bed, admired night scene of the famous road and took photos.

That was the charm of Nanjing Road, I thought then.

I have walked through Nanjing Road many times on my way to the Bund and remember passing by shops using loud speakers to sell jade brocades, tourists speaking different languages and dialects, and young boys hurriedly chucking flyers and cards in my handbag.

But my interest in Nanjing Road started when I found a 1929 map of Shanghai that highlighted the market value of different zones in the city. A blue-toned, T-shaped zone of Nanjing Road all the way to the Bund was the city's most expensive area that year.

Nanjing Road was constructed in 1851 as "Park Lane" — from Bund to the racecourse on today's Henan Road. It was widely called "Ta Maloo" which translates into "Great Horse Road." The Maloo was extended to Zhejiang Road in 1854 and stretched further to Xizang Road in 1862 as the race course was relocated twice — the last one in today's People's Square.

According to *History of Shanghai* that was published by Shanghai Municipal Council in 1921, English missionary Walter Henry Medhurst suggested that "the settlement road names should be made intelligible to the tens of thousands of natives who had crowded into the limits for safety from the Taiping Rebellion." Thereafter, Park Lane was renamed Nanjing Road after the ancient Chinese capital city.

Early last century Nanjing Road was upgraded to a world-class shopping street after Chinese merchants built four modern department stores — concrete structures with modern equipment and high towers — along the street.

"At night they were illuminated by thousands of electric lights, and helped make the upper part of Nanking Road a 'great white way.' Most significant of all, they answer to a demand and are crowded by shoppers, many of whom were visitors to Shanghai from other cities. They mark both the gradual process of the modernization of China and the growing prosperity of Shanghai," F. L. Hawks Pott described in his 1928 book *A Short History of Shanghai*.

The 1934 version of "All About Shanghai A Standard Guidebook" notes this principal retail business street in Shanghai was "designated by an eminent American author as one of 'the seven most interesting streets in the world'."

In 1945, the local government renamed the former Bubbling Well Road to Nanjing Road W. — and the other end became Nanjing Road E. The entire stretch came to be known as Nanjing Road that stretched 5 kilometers. The street became so prominent that it came to symbolize old Shanghai, and nicknamed "Shi Li Yang Chang" or "10-mile-long foreign metropolis".

However it was not an entirely "foreign street". Archive photos show that the eastern section of Nanjing Road, from the Bund till Henan Road, was more foreign-owned and had a Western look, while the street scenes from Henan to Xizang roads were more Chinese. Nanjing Road gradually took its shape after enduring longtime contacts and competitions between Chinese and Westerners on social and urban space.

There was once an analogy narrated by early Shanghai expatriates that if the Bund was like a bow, then Nanjing Road was the arrow, flying westward, which has been the direction that has guided Shanghai's urban development for a long period of time.

In 2017, one year after my book *Shanghai Bund Architecture* was published, Shanghai Huangpu District Government launched a big-scale regeneration of Nanjing Road for a transformation in the new times.

To attract younger customers and revive "yesterday glory", several century-old emporiums was renovated and highlight of the plan was a glass dome and a gallery in the air to beautifully link the 1936 Sun Sun Co. to two new buildings. The fancy Central Arcade invested by Jewish tycoon turned to be a chic landmark named "The Central" for upmarket shops and fine dining. The 1930s Shanghai Power Company and 1980s East China Electrical Power Building converted into a boutique hotel. China's leading technology giant Huawei announced to opened a grand 5000-square-meter store, its largest flagship on earth in the former silk emporium of Laou Kai Fook.

And more excitingly, Shanghai History Museum opened to the public in the former clubhouse of Shanghai racecourse, which had shaped Nanjing Road and had a profound effect on the city's urban development. The new museum not only conserved many historical details such as ornaments shaped like horses' heads but also added modern functions and comforts. Moreover, the history museum displayed many precious antiques about history of Nanjing Road.

At the end of 2019, Nanjing Road embraced another renovation after celebrating 20 years' anniversary of the Pedestrian Street. After this project, the pedestrian part of Nanjing Road will extend from Henan Road M. to Sichuan Road M. The road will also be elevated and paved with identical stones for a better look and feel.

While exploring and researching this historic road, I have survived the dust, dirt and noise and, at times, been forced to wear a safety helmet. But I was fortunate to have witnessed the whole process.

It was such a journey filled with architectural gems, historical stories and devoted spirits. It also gave birth to the newest conceptions and methods of preservation and regeneration. Compared with the Bund, Nanjing Road is East-meets-West but more diversified. Works designed by modern China's first-generation architects graced this road all the way, who had artfully combined Western technology with Chinese elements. Nanjing Road also saw more Chinese owners of buildings than the bund.

Following the direction of Shanghai race horses, I started from the bund and walked westward along the 19th-century Nanjing Road that is undergoing a 21st-century regeneration. It was an extraordinary journey that I'd like to record and share with you.

Michelle Qiao
June 27, 2020
Moon Atelier, Wukang Road

郑时龄的回忆

中国科学院院士 南京路步行街总设计师

Memories of Zheng Shiling, a Academician of the Chinese Academy of Sciences and the chief designer of Nanjing Road Pedestrian Street

1997年南京路曾试行过周末步街，效果很好，所以1998年开始从河南路到西藏路确定为步行街，长度为1033米，宽度是18-28米，宽宽窄窄变化的。

我们提出要调整业态，当时南京路的广告牌像苍蝇拍一样，广告太多灯也太亮，南京路有37个黑色的铸铁窨井盖，都是通信不是下水的。我们改为紫铜来做，把上海的历史一个个做成图案画在上面，如最早的龙华塔。有几个是我自己画的。

南京路上没有排水的窨井，其实排水是有条看不出来的缝。我们把缝做成斜的，看不到底。缝不能太大，否则穿高跟鞋会掉下去，但也不能太窄因为要排水，很是动了一番脑筋。我们曾经担心排水量不够，后来事实证明下大雨南京路没有淹过。南京路步行街有个好处，石头用的是8公分厚的，下面用钢筋混凝土做底，相对比较厚实耐用。

南京路步行街因为下面有地铁，没办法种树，用树池来解决，加了12棵树，就在置地广场那边。

1998年地铁还在施工，南京路项目同时进行，是一半一半做，没有把路封掉而商店还营业。

当时也提了个思想，不应该只是一条路，而是一个网络，不过当时没有条件就先做南京路步行街。未来南京路步行街应该向两边延伸，一直延伸到外滩，像首尔的明洞一样与九江路和宁波路成为网络，往西延伸过Bubbling Well Road (南京西路旧英文名)一直到黄陂路。"

"In 1997, Nanjing Road had carried on a weekend pedestrian street trial and it worked well. So in 1998, it was determined to turn Nanjing Road into a pedestrian street Which is 1033 meters long from Henan Road to Xizang Road, with the width varied from 18 to 28 meters.

We proposed to rearrange businesses of the commercial street. At that time, the billboards along Nanjing Road looked like fly swatters. There were too many billboard lights which were all too bright.

There had been 37 black cast iron manholes for the maintenance of telecommunication without the function of drainage. We replaced them with copper ones and paint the history of Shanghai to decorate each of them, such as the city's earliest Longhua Pagoda. Several of them were drawn by myself.

There was no drainage well on Nanjing Road. In fact, drainage was through seams that could not be detected. We designed unique thin and oblique seams, otherwise ladies would fall off when wearing high heels, but it should not be too narrow for proper drainage. So we crushed our brains during the designing process. We had worried that the drainage volume was not enough, but it turned out that Nanjing Road has never flooded under heavy rain. The materials used are for Nanjing Road Pedestrian Street, such as the 8-centimeter-thick stone brick for its surface and the reinforced concrete for its bottom, are another good point of project, which has made the street relatively solid and durable.

Since there would be a metro line below, planting trees along the pedestrian street was not possible, so we used the tree ponds as a solution. A number of 12 trees were added this way over the Landmark Square.

The subway was still under construction in 1998, and the Nanjing Road project was carried out at the same time.

I had made a proposal that it should not be just one road, but a network. But at that time conditions only allowed for one pedestrian street project. In the future, Nanjing Road Pedestrian Street should extend from both ends, all the way to the Bund, forming a network with Jiujiang Road and Ningbo Road like Myeongdong in Seoul, and expand westward to cross Bubbling Well Road (former English name of Nanjing Road W.) until Huangpi Road. "

开启"摩登时代"

Modern Era Starts Here

1929年8月1日，位于沙逊大厦内的华懋饭店（Cathay Hotel）盛大开业，宾客如云。英文《字林西报》刊登的广告十分诱人，称这座酒店是"艺术与奢华的完美结合"。

广告并未夸大其词。同济大学常青院士研究发现，沙逊大厦是整体受到装饰艺术风格（Art Deco Style）影响的第一栋新建筑。他关于这座大楼的著作就题为《摩登上海的象征》。这种注重装饰、色彩绚丽的新风格源于1925年法国巴黎装饰艺术博览会，在世界各地一度盛行。

沙逊大厦的设计建造过程一波三折。大楼原址是19世纪沙逊洋行所建的"沙逊姊妹楼"，位于上海地价最高的地块——滨江外滩与南京路构成的"T"形黄金地段。

1925年，执掌沙逊洋行的维克多·沙逊（Victor Sassoon）爵士计划在此翻造一座高楼。公和洋行设计的方案是一座中规中矩的新古典主义风格巨厦，于1926年开始施工。

1928年12月22日，美商《密勒氏评论报》（The China Weekly Review）报道透露，工程开工一段时间后方案改变，业主决定在大厦里开设一家豪华酒店。所幸原设计方案经过巧妙调整，得以让地基结构不做激进的改动。

主持和平饭店修缮工程的著名建筑师唐玉恩提到，新方案要求将沙逊大厦的上部设计修改为酒店，东面临江处顶部还要提升两层高度，可以想象如此重大修改的难度。原先A字形的平面很长，横向发展的建筑最后被修改为垂直线条的装饰艺术风格，建筑师的功力可见一斑。修改后的方案在商业上很成功，也给上海这座城市带来独树一帜的建筑形象。

沙逊大厦建成后占地面积4617平方米，总高77米。底层租给银行和货廊，二至四层是写字间，沙逊洋行位于四楼，沙逊爵士的公寓位于顶层十楼。其余楼层都是华懋饭店，八层有大酒吧、舞厅和中式餐厅，九层有夜总会和小餐厅。

沙逊大厦呈现装饰艺术化的

浪漫新古典风格，不仅成为外滩和南京路的地标，更把上海全面推向了摩登的装饰艺术时代。短短数年后，国际饭店和百老汇大厦等装饰艺术风的大楼纷纷落成。

1928年的报道提到，上海公众对于这座新厦内的豪华酒店充满期待。华懋饭店将是一家设施一流的酒店，主要设备和家具装饰都从英国进口，有200间客房和精美餐厅。

1929年饭店开业后，果然没有辜负众人的期待，被美商《大陆报》誉为远东地区最好的"宫殿般的酒店"。饭店著名的九国套房以中、英、美、印等国主题布置，异国情调浓郁。每个套房都含有卧室、餐厅、休息室和两间浴室，并配有嵌入式壁橱和大理石浴缸。银质龙头流淌出净化水。舞厅和宴会厅镶嵌着价值连城的法国"拉立克玻璃"。这种由法国玻璃艺术家拉立克先生（Rene Lalique）设计的乳白色玻璃经灯光照射即可闪现蓝光或耀眼的桔红光芒，美丽无比。"拉立克玻璃"后来也成为装饰艺术风格的经典手段。

华懋饭店开张后，接待了许多政要和社会名流，如卓别林、萧伯纳、马歇尔将军和斯诺等。著名剧作家科沃德（Noel Coward）还在酒店里写出名作《私人生活》。

1935年，后来以《宋氏姐妹》一书出名的美国女作家项美丽（Emily Hahn）到上海后，曾是这间梦幻酒店的常客。在自传《我的中国》中，她深情回忆那一段充实快乐的上海生活：

"如果不去弗雷兹夫人家作客，我也许就约一位女朋友到华懋饭店吃午餐。我们习惯到大堂先喝一杯，看看能否遇到有趣的男士。"

事实上建造大楼的犹太富商维克多·沙逊爵士（Sir Victor Sassoon）就是一位相当有趣的男士，后来他成为项美丽人生导师般的密友。爵士来自号称"东方罗斯柴尔德"的巴格达沙逊家族，极富商业天赋，和平饭店一楼的甜品店就以他命名。

爵士喜欢在自己位于大厦顶层的公寓举办主题奇异的派对。

他爱好广泛，深爱摄影，项美丽姐妹一张颇为传神的侧影就出自他手。

饭店也见证了很多重要的历史时刻。1998年10月14日，中国海协会会长汪道涵和台湾海基会会长辜振甫继五年前在新加坡的首次"汪辜会谈"后，在和平饭店"和平厅"再度握手，进行第二次"汪辜会谈"，达成四点共识，对两岸关系发展带来了积极影响。汪道涵还在"九霄厅"宴请辜振甫一行，共同品尝了阳澄湖大闸蟹。

时光流转到2007年，已成为和平饭店北楼多年的沙逊大厦为迎接2010上海世博会，迎来长达三年的整修。大修前，常青院士的团队对饭店进行了80年来首次全面测绘，这座著名历史建筑给他留下了"一个老贵族"的印象。

"他是位老人，但他是个贵族，一个'lord'，非常讲究，非常正统。"常青回忆道。

而对于主持修缮工程的建筑师唐玉恩来说，恢复酒店大堂八角中庭和丰字廊的历史原貌富有挑战。

这是因为20世纪50年代后，和平饭店在收归国有时归属旅游、电信和商业三家单位，形成了底层大堂由和平饭店、外贸商场和电信局分割使用的局面。八角中庭及其南廊都是外贸商场，还搭建了钢筋混凝土夹层及螺旋楼梯。

唐玉恩回忆道，这里楼下卖衬衫羊毛衫，楼上还是衬衫羊毛衫。她曾以为原来这个建筑就是这样，但历史图纸和照片告诉她空间不是这样的。

大修前外贸商场搬离，混凝土楼板和楼梯拆除，呈现出一个巨大的空间，大堂的丰字廊得以贯通，但八角中庭的玻璃顶仍旧黯淡。这是由于沙逊大厦初建时还没有夹胶玻璃，为了防止普通白玻璃碎落，两层玻璃天窗内外设有两层铁丝密网来防护。后来铁丝网锈蚀加上玻璃陈旧，八角中庭就暗了下来。2007年大修工程时，唐玉恩的团队使用碎裂也不会滑落的暖棕色夹胶玻璃，玻璃顶就不需要加设铁丝网了。

"铁丝网一拆掉，八角中庭

的亮度马上不一样了，就像初建成时一样明亮。饭店的老员工都很高兴，说这是'重铸辉煌'，有人感动得流下了泪水。"唐玉恩回忆道。

而在修缮和平饭店标志性的绿色四方锥顶时，因为没有彩色历史照片，修缮团队推测屋顶原本是略深的紫铜皮颜色，而绿色是铜板氧化的结果。不过，由于自20世纪50年代以来，人们对这个外滩屋顶的城市记忆一直都是绿色，大修时保留了屋顶的深绿色。

今日走进和平饭店，好像走进一个古旧的澄黄色的梦。在重现辉煌的八角中庭，大穹顶和紫铜吊灯都笼罩着半透明的暖黄玻璃。昏黄的光影中，各色几何图案若隐若现，与酒店里弥散的氤氲香气裹挟在一起，让人心潮澎

湃，恍如隔世。

2019年8月1日晚，和平饭店举行了隆重的90周年庆典，也是宾客如云。精心设计的庆典带有浓郁的摩登时代元素，由老年爵士乐队演奏的老歌助兴，穿闪亮黑裙、头戴白色羽毛的侍者招待着宾客。一切，都宛若1929年8月1日那个开启"摩登时代"的夜晚。

昨天：沙逊大厦　**今天**：和平饭店　**地址**：南京东路 20 号
建造时间：1929 年　**建筑师**：公和洋行　**建筑风格**：装饰艺术风格
参观指南：酒店的和平收藏馆有很多旧照片和老家具，值得一看，或到一楼甜品店"Victor's"喝杯咖啡。最好选择坐在窗边，沐浴在清晨阳光之下看南京路街景。

An inviting advertisement announced the grand opening of the Cathay Hotel in the North China Herald in 1929, noting it was a "wonderful combination of art and luxury".

The advertisement was honest. Tongji University professor Chang Qing, whose team surveyed No. 20 before its 2007 renovation, discovered the building "was China's first Art Deco building but the original plan was a neo-classical high-rise".This decorative modern style was first dated in the 1925 Exposition in Paris, which soon gained popularity around the world.

The design and construction of Sassoon House went through many twists and turns. The site was on the Sassoon sister buildings in the 19th Century, located on Shanghai's most expensive plot--a T-shaped area comprised of partial Bund and Nanjing Road E. The original plan was a classic, horizontal high-rise office building named the Sassoon House. But in 1926, the owner, Jewish tycoon Victor Sassoon, decided to convert the upper part of the building into a luxurious hotel, elevating the tower top for commercial reasons.

On December 22, 1928, *the China Weekly Review* revealed that the scheme was changed after the construction work was kicked off. The owner decided to establish a luxury hotel in the building. Fortunately, clever adjustments to the original design avoided radical changes to the structure of the building foundation.

"You can't add the architectural load at will after the foundations were all planted. George Wilson from Palmer & Turner skillfully revised the horizontal plan to an imposing, even taller vertical structure in an art-deco style with a striking tower. As an architect, I'd say such a radical change was an incredibly challenging job that highlighted Wilson's superb skills," says renowned architect Tang Yu'en who oversaw the hotel's renovation project from 2007 to 2010.

The modified scheme was not only commercially successful, but also brought a unique architectural image to Shanghai.

Upon completion, Sassoon House covered an area of 4,617 square meters and boasted a height of 77 meters. The ground floor was rent to banks and vendors, while the first to the third floors were offices for Sassoon's company and other enterprises.

The Cathay Hotel occupied the ground floor and the fourth to the ninth floors. Sassoon's penthouse occupied the 10th and 11th floors.

Divided by classic three sections, Sassoon House is not a pure Art Deco architecture but Wilson uses a lot of Art Deco manners, including opalescent Lalique glass. The building also showcases commercial Gothic features, which interestingly echo with the neighboring Club Concordia (demolished in the 1930s to build the Bank of China) in Gothic revival style. Named a "Shanghai Deco" building by professor Chang Qing, Sassoon House brought a modern era to the city. High-rises in Art Deco style including the Park Hotel and the Broadway Mansions were completed within several years.

A 1928 report mentioned that Shanghai public was full of expectation for this new building, particularly the Cathay Hotel. The report stated that the Cathay Hotel would be a top hotel featuring 200 rooms and fine diners with major facilities and furniture imported from UK.

After opening in 1929, the hotel met everyone's expectations. The China Press gave it the accolade of "Palatial Hotel in Sassoon House slated finest in the Far East" .The nine famous "themed" suites, each decorated in a distinctive national style, including Chinese, Indian and English. Each suite had built-in wardrobes. The bathrooms contained marble baths with silver taps and purified water. The hotel's dining rooms were decorated with colorful, blazing Lalique chandeliers. The milky-white glass designed by French glass artist René Lalique would shine a blue light or dazzling orange. It was incomparably beautiful. Lalique glass later became a classic method of art deco style.

After the opening of Cathay Hotel, it hosted many politicians and social celebrities such as Charlie Chaplin, George Bernard Shaw, General George Marshall and Edgar Snow. The famous dramatist Noel Coward wrote his celebrated work *Private Lives* here.

The hotel also witnessed many important moments in history. On October 14, 1998, five years after the first Wang-Koo Talks in Singapore, Wang Daohan, president of the Marine Association, met with Koo Chen-fu, chairman of the Taiwan's Straits Exchange Foundation. They shook hands at the "Peace Hall" of the hotel, con-

ducting the second "Wang-Koo Talks".

Back to 2007, Sassoon House, which had been the Peace Hotel's Northern Building for many years, underwent a three-year renovation in preparation for the 2010 Shanghai Expo.

When architect Tang was commissioned to restore this gigantic landmark, she was faced with an equally challenging job compared with Wilson's some 80 years ago.

Social change in China has split the history of the legendary Sassoon House into two parts. As the Cathay Hotel, it was famous throughout Far East and set a precedent for luxury and glamor. But Sassoon's parties came to an end with the outbreak of the World War II. He left Shanghai in 1941.

The building was used by the Shanghai Municipal Government after 1949. It reopened later as the state-owned Peace Hotel in 1956 but the ground floor was shared by the hotel, East China Telecom Bureau and Shanghai Commercial Bureau which then had a clothing shop in the rotunda. In the early 1990s, the latter added a mezzanine and a spiral staircase in the shop for flooded shoppers on Nanjing Road, which entirely changed the look of the historical rotunda.

It took some work to persuade the three state-owned bureaus to cooperate, open separation doors and connect the rotunda and the

THE NEW SASSOON HOUSE
SHANGHAI

arcade again, like it was in 1929.

The highlight of the project was to restore the rotunda, which looked rather dim then, to its glory of yesteryears.

"In the 1920s, glass was still a novel material whose quality was not as good as today. So two layers of iron nets were designed above and under the glass dome of the rotunda to eliminate injuries in case a piece of broken glass fell down," Tang explains.

"Now we have laminated glass. Two pieces of glasses stick together with highly elastic glue, which won't fall down if it snaps, so the iron nets are not necessary. The moment we removed the rusting iron nets and replaced the old glasses with new ones, the rotunda lit up and looked what it did in 1929. The hotel's old employees were so happy and they said we rebuilt yesterday's glory," she adds.

Today, Walking into the Fairmont Peace Hotel is like walking into a nostalgic dream. In the resplendent octagonal rotunda, the large dome and copper chandeliers are shrouded with translucent warm yellow glass. In the dim light, geometric patterns in various colors loom, in harmony with dense aroma that pervades the hotel. It has created a surge of emotions as if one has been cut off from the outside world for generations.

After a three-year renovation, the building reopened in 2010 as the Fairmont Peace Hotel, which hosted a grand 90 years' celebration on August 1, 2019. The celebration gala attended by hundreds of guests was designed in Art Deco style highlighted by old jazz songs, which mirrored the opening night of this "hotel of art and luxury" in many ways.

Yesterday: The Sassoon House   **Today:** Fairmont Peace Hotel
Address: 20 Nanjing Rd.E.   **Date of construction:** 1929   **Architect:** Palmer & Turner
Architectural style: Commercial Gothic & Art Deco
Tips: The Peace Gallery exhibits some historic archives of the hotel. I'd also suggest have a drink at window-side table of the "Victor's' café with a view of the Nanjing Road and usually with nice morning sunlight.

斯诺的沙逊套房之夜
Edgar Snow's night in the Sassoon Suite

1949年，著名美国记者、《西行漫记》作者斯诺与妻子海伦的婚姻结束了，他们曾在南京路相识相恋、在外滩求婚。1960年，已再婚的斯诺带着过去的经历访华，是第一个回到中国的外国记者。斯诺入住了上海和平饭店，酒店为他提供了沙逊套房。斯诺本来觉得一个单间就够了，但是好奇地想看看沙逊爵士的房间。

"套房位于塔楼，高居江面之上，类似都铎风格。墙壁用镶木墙裙装饰得很好看，宽大的格窗引进充足的光线。卧室很奢华，有两张巨大的床，分别附有两间很大的砌了瓷砖的浴室。此外，还有一间面积跟小公寓差不多的化妆间，一个专属门厅，仆人房间、酒吧、厨房和备餐室，带壁炉的私人餐厅，一间宽敞的起居室，同样带有壁炉。"斯诺在《今日红色中国》一书中回忆道。

这个套房曾经属于建造沙逊大厦的犹太富商沙逊。为了体验沙逊爵士早晨在外滩醒来的感觉，斯诺决定在这间套房住上一晚。不过，这一晚他没有休息好。他从摩登大厦之巅俯瞰江景，一幕幕往事在心里汹涌澎湃。他想起年轻的自己在这座城市度过的青春岁月，"坠入爱河、陷入一段短暂又经历丰富的婚姻、见证了两场战争……"

The marriage between famous American journalist Edgar Snow, author of *Red Star Over China* and Helen Foster ended in 1949. In 1960, Snow, who had remarried, visited the People's Republic of China as the first American with the former experience he had with old Shanghai. He was also the first foreign correspondent to return to China. Snow

checked into the Shanghai Peace Hotel which offered him Sassoon's former suite. Thinking that a single room to be sufficient, Snow was curious about the suite.

"The suite was at a top tower, high above the river and in a Tudor style. Wood paneling was used on the walls and the decoration on the wainscoting was very pretty. The wide lattice windows brought in sufficient light. The bedroom was very luxurious with two enormous beds and two en-suite tiled bathrooms. In addition, there was a dressing room the size of a small apartment, an exclusive hallway, a servant's room, a bar, kitchen and a pantry, as well as a private dining room with a fireplace, and a spacious sitting room which had a fireplace." recalled Snow in his book *Red China Today*.

The suite formerly belonged to Jewish tycoon Victor Sassoon. To experience the feeling when Sir Sassoon woke up in the morning, Snow decided to stay in the suite for one night. However he did not have good rest that night. Enjoying a bird's view of river scene from the top of this modern edifice, memories surged in his heart. The days he spent in this city when he was young occurred to him, falling in love, into a short but eventful marriage and witnessing two wars...

南京路的"王宫"

The"Palace" on Nanjing Road

在上海的城市记忆中，位于南京路外滩转角的红色酒店是和平饭店南楼，其实它的历史比和平饭店更加悠久。

大楼原为19世纪中期所建的中央饭店，高仅三层，1906年开工翻建为6层高的汇中饭店。值翻建之际，董事会决定将酒店的英文名由"the Central Hotel"改为"the Palace Hotel"。

这个新名字名副其实。大楼由司各特设计，清水红白砖的立面生动夺目。远远望去，酒店好像一座漂浮在外滩的童话王宫，而"The Palace Hotel"确有"王宫"的意思。

6层高的酒店有120间客房和一间200人大宴会厅，作为当时的"高层建筑"还安装了2部我国最早的电梯。密布繁复雕花的南京路入口有一扇老式转门。

中国近代史上许多重大事件就发生在汇中饭店。1909年2月，第一次国际禁毒会议召开，来自中、美、英、法、德、俄、日等13国代表在汇中饭店商讨禁除鸦片事宜，史称"万国禁烟会"。会议通过九项决议案，赞成中国政府的禁烟条令，也推动了第一部国际禁毒公约《海牙鸦片公约》的缔结。1911年辛亥革命成功后，12月25日上海各界在汇中饭店欢迎孙中山先生就任南京临时政府大总统，孙中山在会上发表演说。1927年，蒋介石和宋美龄在此订婚。

此外，汇中饭店因为有上海最早的屋顶花园而风靡一时。花园由人工草坪和绿色藤蔓装饰，有一个巴洛克式的亭子。在这里边喝咖啡边欣赏江景和浦东田园风光，是绝佳的享受。

可惜1912年8月的一场大火影响了酒店生意，外白渡桥边的礼查饭店新楼又吸引走不少客人。后来汇中饭店曾一度复兴，但更摩登的华懋饭店（今和平饭店）和国际饭店也相继建成。

1933年6月，曾在汇中饭店担任7年总经理的博思（Frederick Boss）接受报纸采访，提到上海的酒店客人是全世界国际化程度最高的，这一点连伦敦和巴黎都无法与之相比。

"上海的酒店很现代，服务效率高，非常舒适。上海每家大

型酒店都有现代卫生设施，中央供暖，和空气新鲜的带浴室的客房。每天，平均每位客人在上海的酒店消费12到15块美金。"博思说道。

抗日战争期间，酒店曾被日军占据，新中国成立后又用作办公，直到1965年改为和平饭店南楼，人们几乎忘却了它昔日的荣耀。

到了20世纪80年代，"王宫"又差点被拆除。2010年，"19号"终于被改造为斯沃琪和平饭店艺术中心，包含名表店和艺术套房，迎来新的命运。

改造工程保留了大木梯和蒋宋订婚厅的原貌。展览厅裸露着老建筑的原始立柱和灰色砖墙，细心观察能看出百年前客房并不宽敞的开间布局。

如今，酒店顶楼两层是主题套房和高级餐厅，中部两层用于由斯沃琪集团首席执行官海耶克（Nick Hayek）发起的艺术项目——提供18套工作室，供世界各地受邀的艺术家进驻创作。很多艺术家留下了带有上海建筑灵感的作品。

曾经参加项目的西班牙画家白怀义（Juan Antonio Banos）坦言最大的灵感源泉是南京路喧嚣的夜晚。偌大而安静的画室里，他常常从这"王宫"的券窗向外眺望，那么多闪光灯，还有夜上海的迷离灯光，后来都凝固在他斑斓的画作中。

昨天：汇中饭店　**今天**：斯沃琪和平饭店艺术中心　**地址**：南京东路23号
建成时间：1908年　**建筑师**：司各特（Walter Scott）　**建筑风格**：维多利亚殖民地风格
参观指南：艺术中心部分对外营业。建议从古老的楼梯拾级而上，到展览厅欣赏裸露的原始墙壁和柱子，想象昔日客房的格局空间。

Widely known for decades as the Peace Hotel South Building, the Palace Hotel faced with red-and-white bricks has a much longer history than the Peace Hotel that opened in 1929 as Cathay Hotel. Designed by British architect Walter Scott and opened in 1908, the hotel was constructed on the site of the even older Central Hotel, which was built in 1875.

When renovation plans for the hotel were released, the hotel board was so thrilled that they decided to change the name, the Central Hotel, to the Palace Hotel.

The hotel lived up to its new name and did look like a palace straight out of fairy tales. A variety of different window shapes were created for different floors, including semi-circular arches, diminished arches and pointed arches. The grand entrance on Nanjing Road E. had exquisite carvings and a revolving door.

As one of the largest, tallest and most spacious hotels in China, the Palace Hotel had 120 rooms, spacious dining area and banquet hall besides a famous roof garden. It had the city's first elevator, an Otis elevator.

It was the venue for the 1909 International Opium Commission meeting, marking the first step against opium trade. In 1911, Dr. Sun Yat-sen hosted a banquet here to celebrate his victory in the presidential election. Kuomintang leader Chiang Kai-shek and his wife Soong Mei-ling held their engagement reception on the top floor in 1927.

In its prime, the hotel was also known for its elaborate roof garden, with a Baroque tower, a pair of cupolas and artificial lawn. Green vines wound around the railings. It was such an enjoyment to sip a glass of whisky on ice while admiring the music from the Municipal Orchestra concerts in the Public Garden on summer weekends, or drink hot coffee under warm sunlight while appreciating the river scene on a winter afternoon.

Unfortunately, a fire on the rooftop on August 15, 1912 changed the fate of the hotel. According to the North China Daily News, the fire broke out in the northwest corner of the roof in the ornamental cupola and raged for more than an hour.

After the fire the hotel lost its clientele to a new building nearby — the Astor House Hotel. In the late 1910s, the Palace Hotel had some good times but faced strong competition from modern "skyscraper" hotels in the 1920s, such as the neighboring Cathay Hotel and the Park Hotel, both on Nanjing Road.

Frederick Boss, manager of the former Palace Hotel, said in June 1933 that "Shanghai hotel guests are the most cosmopolitan in the world."

"People of all walks of life, all nationalities, all creeds, pass through Shanghai. Nowhere, not even in London or Paris, is there such a distinctly international hotel patronage," Boss, who until then had worked in the hotel for seven years, said during a newspaper interview.

"Shanghai's hotels are second to none in the world. They are modern, efficient and comfortable. Modern sanitation, central heating and airy rooms with baths are to be found in every large Shanghai hotel. The average guest who puts up at a Shanghai hotel pays from US$12 to US$15 for a room with meals," the hotel manager said.

According to the Huangpu District Archives, the building was occupied by the Japanese during World War II and used by several state-owned organizations after 1949. In 1965, it reopened as the Peace Hotel South Building. This history-rich hotel lost some of its shine after becoming the South Building of the Peace Hotel.

The past glory of the Palace Hotel was almost forgotten, but a renovation in 2010 transformed this historical building into the new-concept Swatch Art Peace Hotel.

The 2010 renovation changed the layout but retained the dark-toned wooden staircase. The original interior gray-brick walls and

columns are exposed. The famous roof garden opens for several months in a year. The building still bears the inscription "1906" above the main entrance on Nanjing Road -- 1906 being the scheduled completion date.

Today, the ground and first floors feature flagship shops of the Swatch Group and a spacious hall showcasing exhibitions. The top three floors were turned into stylish hotel suites and chic restaurants.

In between dazzling watches and fancy suites, the middle two floors contribute to an "Artist in Residence"program. Each year, as many as 40-50 artists from around the world are offered a stay for months free of charge. Each receives a room and separate studio. Many of them created artistic works inspired by Shanghai architecture.

Spanish contemporary artist Juan Antonio Banos, one of the resident artists, said the flashes of cameras on the Nanjing Road just outside the window and all the lights of nighttime Shanghai gave him endless inspiration.

Yesterday: The Palace Hotel   **Present:** The Swatch Art Peace Hotel
Address: 23 Nanjing Road E.   **Date of construction:** In 1908   **Architect:** Walter Scott
Architectural style: Victorian colonial
Tips: The hotel is partially open to the public. I would recommend climbing the antique staircase to appreciate the exhibition room featuring exposed original walls and columns.

嘉陵大楼与罗迦陵

Liza Hardoon Building and Mrs. Hardoon

南京东路99号嘉陵大楼原名迦陵大楼，是犹太富商哈同夫人——罗迦陵的产业。1936年迦陵大楼落成在即，英文《大陆报》称这座6层高的大厦是"市中心又一座壮观的建筑"。

"从街道上看，大楼外观相当现代。外立面大部分的墙面由人造石覆盖，拱肩墙以面砖砌就，用水泥固定于混凝土墙上。建筑整体是混凝土结构，还使用了防火材料。" 报道写道。

设计大楼的建筑师德利（Percy Tilley）告诉《大陆报》（the China Press），新大楼融商铺、写字楼和公寓为一体，租赁情况非常好：

"因为人们对于市中心宜居空间的需求，大楼里的公寓房很受欢迎。整个顶层都是供居住的公寓，提供带浴室的房间。大楼通往写字楼和公寓的主入口在南京路上，有一个宽阔的楼梯和两台电梯。"

因为当时建筑技术的发展和简洁的设计，大楼的建造仅用了一年多时间，于1937年12月竣工。大楼外观采用垂直竖向线条，局部有几何装饰。灰色大楼装有钢门和钢窗，内部装修使用了进口的桃木和柳桉。

在老上海，大楼里最著名的商户要数美商大通银行（Chase Bank）。今天，大楼仍然作为工商银行的一家支行在繁华的南京路营业。室内的历史细节虽已很少，但银行把嘉陵大楼和南京路的黑白老照片放大，挂在营业厅的墙上展示。

嘉陵大楼因罗迦陵而得名。大楼落成前5年，她的先生、犹太富商哈同去世，罗迦陵继承了数额巨大的遗产，成为那个时代的亚洲女首富。

罗迦陵1864年出生于上海一个贫苦的家庭，1886年嫁给哈同。1868年从巴格达来到上海闯荡的哈同同样出身贫苦。他身无分文，在老沙逊洋行从最底层的"司阍"（看门人）干起，因为勤勉又有头脑，很快掌握了上海地产投资的诀窍。1886年，哈同加入新沙逊洋行主管房地产部，1901年在南京路成立哈同洋行从事贸易和房地产投资，后来成为上海滩著名的地产大王。

嘉陵大楼所在的位置原是老牌英资百货——福利商店（Hall & Holtz）。在哈同家族数不清的地产投资项目里，设计简洁的嘉陵大楼是少数现代风格的建筑之一。

研究上海犹太人历史的上海社科院王健教授介绍，哈同曾在原公共租界工部局和法租界公董局都担任过董事，这是不多见的。

"这也是因为哈同对于城市发展独到的眼光。他出任新沙逊洋行主管房地产的大班协理后，仔细研究分析了上海房地产未来的发展走向，决定将投资的重点放在南京路。当时西藏路一带还是郊外，是他建议公共租界应该向西发展，直到今天静安寺一带的位置。当时其他人也有向北和向南发展的建议。"王健说。

独具慧眼的哈同认为："南京路居虹口、南市之中，西接静安寺，东达黄浦，揽其形胜，实为全市枢纽，其繁盛必为沪滨冠。"因此，他不断低价购入南京路的地产。后来正如他所料，南京路果然成为上海最繁华的商业街，而哈同的财富也随着南京路的快速发展不断地增值。

王健教授统计，哈同1931年6月19日病逝时拥有16块南京路黄金地段的地产，面积达111.578亩，占南京路地产总面积的44.23%，几乎拥有了半条南京路。

哈同迎娶罗迦陵后，在上海安家立业。妻子对哈同的影响很大，她劝说哈同信奉佛教，赞助佛教书籍出版，创办教育机构，喜爱上中国文化。没有子嗣的哈同夫妇领养了十多个中外孤儿。

1931年哈同去世后，英文《字林西报》刊文写道，"这对夫妇在一起快乐地生活了45年，互相奉献。哈同对罗迦陵敬意有加，总是称她为'我的妻子'。"

昨天：迦陵大楼　**今天**：嘉陵大楼　**地址**：南京东路 99 号

建造时间：1936 年　**建筑师**：德利洋行 Percy Tilley

参观指南：大楼底层为工商银行营业厅，内部历史细节不多，可以欣赏历史照片。大楼顶部的几何装饰十分别致。

The Liza Hardoon Building at 99 Nanjing Road E. was named after a Chinese woman who inherited her husband's estate in 1931 and became Asia's wealthiest woman. Today, the edifice is called Jialing Building — from Luo Jialing, the Chinese name of Liza Hardoon.

Born in 1864 to a poor family in Shanghai, Luo married Jewish tycoon Silas Aaron Hardoon in 1886. Hardoon was penniless when he arrived in the city from Baghdad in 1868. He found work with David Sassoon, Sons & Company as a watchman but quickly mastered the art of acquisition of commercial property, chiefly retail property, along Shanghai's main thoroughfare.

When Jialing Building was near completion in 1936, *The China Press* called the six-story structure "one more imposing building in the Central District" .The building housed shops, business offices and residential quarters.

"Viewed from the streets, the building has a modern appearance. The greater proportion of the exterior wall space is done in cast stone blocks, while the spandrel walls are in face brick cemented to the concrete walls. The whole structure is of reinforced concrete and other fire-proof materials," the report said.

The newspaper also noted that all the shops in the new building had been rented to good tenants, according to architect Percy Tilley.

"It is said that the apartment section is popular due to the demand for comfortable quarters in the downtown area. The apartments occupy the whole of the top floor. About 50 bed-sitting rooms with baths have been provided. The main entrance to the business offices and the apartments is on the Nanking Road side of the building. The upper floors are reached by a broad stairway. Two

lifts are also provided," the report said.

The simple-cut Jialing Building was one of the few modern constructions among the former properties of the Hardoon Family. The construction work of the building was completed in a little more than a year due to advanced architectural technology in the 1930s and the simple design.

The building was designed with a stepped massing along Sichuan Road M. The façade, which has vertical lines, was treated in such a simple-cut way that only the top part is adorned with some continuous geometrical patterns. The interior decoration applies peach timber and imported lauan.

In old Shanghai, the building's famous tenant was Chase Bank. It also houses a bank today, a branch of the Industrial and Commercial Bank of China. The interior has changed a lot but the bank has hung some archival photos of the building and Nanjing Road to showcase the past.

"Hardoon was the only man in the history of Shanghai who had served as a member of the board for both Shanghai Municipal Council (that ruled the international settlement) and French Municipal Council (that governed the French concession)," says Wang Jian, a professor from the Shanghai Academy of Social Sciences, who authored "Shanghai Jewish Cultural Map".

"That was due to his vision of the city's development. At that time Xizang Road was still a suburban area and it was him who suggested the international settlement should expand toward the west, till today's Jing'an Temple. There were other opinions like extending to the north or the south," he adds.

Hardoon himself profited from his vision. He founded his own company on Nanjing Road and made a huge profit by investing in the city's booming real estate, especially "Shanghai's Fifth Avenue," today's Nanjing Road.

"After marrying Luo Jialing, Hardoon, a lonely expatriate, finally had a home in Shanghai," Wang says, adding Mrs Hardoon had a big influence on her husband.

"She influenced Hardoon to believe in Buddha, pay for the publication of books on Buddhism, found a college and love Chinese culture. Without kids, they adopted nearly 20 Western and Chinese orphans. One of Hardoon's Western adopted daughters told me that the Hardoons appreciated Peking Opera and used chopsticks at home. The daughter was fluent in both Mandarin and Shanghai dialect," Wang says.

The couple's former home, the beautiful 26-acre Aili Garden, where the Shanghai Exhibition Center stands now on Nanjing Road W., was a gift by Hardoon to his wife. Though the garden is

long gone, several buildings along Nanjing Road such as the Liza Hardoon Building still exist to keep alive the legacy of the Hardoons.

After Hardoon died in 1931 the North-China Daily News reported "the couple lived together happily for 45 years and were devoted to each other. Hardoon treated Luo with respect and always spoke of her as 'my wife'."

Yesterday: Liza Hardoon Building   **Present:** Jialing Building, ICBC Bank
Address: 99 Nanjing Road E.
Built in 1936 to 1937   **Architect:** Percy Tilley   **Architectural Style:** Art Deco
Tips: The building opens during office hours of the ICBC Bank. Please note historical photos of Nanjing Road and the building on the wall.

哈同夫人病逝

Mrs. S. A. Hardoon Dies After Brief Illness

哈同夫人罗迦陵现年78岁,是已故哈同先生中国籍的夫人。她昨晚突发疾病,不久于6点在静安寺路1273号家中去世。

她是远东地区最富有的人之一,其丈夫于1931年6月17日去世后给她留下估计有7亿元的遗产。哈同夫人去世后留下11个养子女。

虽然哈同夫人在上海和其他地方拥有巨大的资源和利益,她生命中最后10年时光都是在家中退休的状态度过的。她的豪宅占据了静安寺路(今南京西路)、福煦路(今延安中路)和哈同路(今铜仁路)的大部分土地。

哈同夫人的去世令她遍及全世界的众多同事友人深感震惊。她是最后一个见证哈同产业在上海崛起过程的人。

出于对中国文化的浓厚兴趣,哈同夫妇曾联合创办了一家名为仓圣明智大学的教育机构。有几十名中国学生受惠于她的帮助在此学习。

哈同夫人的关注点也不局限于中国。1927年,她与丈夫哈同一起敞开大门招待到访上海的英国国防军,为他们提供食宿所需。

南京路及周边地区大部分都是她的产业。按照目前汇率估算,哈同夫人拥有的上海土地价值达7亿元,相当于1千万英镑。

罗迦陵婚后信奉犹太教,她的养子女们也是。

众所周知,公共租界和法租界的大多数主要地产都是哈同家族的财产。几年前,哈同洋行在南京路建造了一座迦陵大楼以资纪念。

哈同夫人的葬礼预计在未来几天内进行。据信,她将被安葬在自家花园里丈夫哈同的旁边。

Mrs. S A. (Liza) Hardoon, 78-year-old Chinese born wife of the late Mr. Silas Aaron, Hardoon, died suddenly in her home, 1273 Bubbling Well Road, at 6 o'clock yesterday evening after a brief illness.

One of the richest persons in the Far East, Mrs. Hardoon was left what is today estimated as a fortune of $700,000,000 when her husband died on June 17, 1931. She is survived by eleven adopted sons and daughters.

Despite her enormous resources and the huge interests which she represented in Shanghai and elsewhere, Mrs. Hardoon spent the last decade of her life in complete retirement in her huge estate which takes in a great part of Bubbling Well Road, Avenue Foch and Hardoon Road.

Her passing will come as a profound shock to thousands of friends and associates throughout the world especially in view of the fact that she was the last remaining figure, who saw the great Hardoon estate rise to power in Shanghai.

Mrs. Hardoon, in cooperation with her late husband, established an institute of education named St. Ching's College in pursuance of her deep interest in all things pertaining to China and her people. Scores of Chinese youths own their education to her assistance.

That she had a heart also for things which were not of China may be shown by her spontaneous action when together with her late husband she threw open their spacious grounds to the British Defence Force which arrived here in 1927 and provided them with everything necessary in the way of food and accommodation.

Nanking Road and the neighborhood were hers in great part and in much more than name alone. It has been estimated that her assets in such local lands should, at present rates, add up to a sum of CN$700,000,000 or an equivalent of about £10,000,000.

She adopted the Hebrew religion after her marriage. All her adopted children follow the Hebrew religion.

For years it has been a well-known fact that most of the important pieces of land in the International Settlement and French Concession were the property of the Hardoon family. A befitting memorial in the shape of the Liza Building on Nanking Road was erected several years ago.

Funeral services for the late Mrs. Hardoon are expected to take place within the next few days. It is believed that she will be buried next to her husband in the garden of their residence.

摘自《北华捷报》1941 年 10 月 8 日
Excerpt from *The North-China Herald*, on October 8, 1941

惠罗公司的沉浮

Ups and Downs of Huiluo Company

南京路上，惠罗公司还保留着百年前的名字，却让人很难看出它的悠久历史。一个世纪里，这座风光一时的英商百货大楼两次浴火，几经沉浮。

南京路外滩段曾有四大外资百货公司——惠罗、泰兴、福利和汇司公司。1882年，惠罗公司创立于印度加尔各答，总部设在英国伦敦。1904年，惠罗公司在上海南京路开店，秉承"质地、价值、服务"的理念，经营百货和进口贸易。

1906年，公司在南京路四川中路转角处建起一座五层高的大楼，同济大学常青院士称这是"南京路的第一幢高层建筑"。历史照片显示，老惠罗大楼是一座设计精美的古典主义建筑，饰有巨柱拱廊和巴洛克山花。惠罗公司经营底部两层，其余楼层用于租赁。由于生意好，惠罗后来收回所有楼面营业。

1921年，惠罗公司主席威尔金森（Mence Wilkinson）访沪，他接受了上海英文报纸的采访。

"惠罗先生缔造杰出商业的办法只有一个：就是保证交易公平并慷慨待人。"威尔金森主席对《上海泰晤士报》（*The Shanghai Times*）记者谈道。

他举了个例子："有一天，我们收到一位内地农场主来信，要求订购一打袜子。这位先生希望可以买到比上次质量更好的袜子，我们就免费给他寄了一打袜子。"这位惠罗公司掌门人认为慷慨大方终有回报，而公司的快速发展也证明这种经营理念是非常英明。

威尔金森先生到访的1921年，正是惠罗公司在上海业务发展的"黄金时代"。当时民族工商业尚未崛起，国际金融市场上英镑贬值，金贱银贵，而在华销售主要收取银元，正利于做进口生意。惠罗公司积累资本后于1929年开始翻建改造南京路大楼，期待"让这座知名商厦与欧洲最前卫的建筑媲美"。

1930年8月，另一家英文报纸——《北华捷报》（*The North-China Herald*）报道了惠罗大楼改造方案，并配了一张匈牙利建筑师鸿达（C.H.Gonda）绘制的效果图。

新方案的建筑风格与位于今日外滩源的光陆大楼有几分相似。两座大楼都是邬达克东欧老乡鸿达的作品。与邬氏作品复杂多样的建筑风格不同，鸿达的建筑趋向更现代的摩登、简洁风格，如外滩14号原交通银行大楼、南京路原新新百货和淮海路上的国泰电影院。

鸿达用他擅长的摩登风格来改造古典精致的惠罗大楼。报道提到，建筑的外立面会彻底改造，去除过时的石雕装饰，设计为在国外已经广泛使用的现代线条。建筑师计划打开墙面空间，最大限度地引入阳光和自然风，在墙上安装现代铰链钢窗。总而言之，这个被重塑建筑的设计关键词，就是"简洁、高雅和有效"。

2019年匈牙利出版的《鸿达》一书里提到，为惠罗大楼的设计呈现了现代气息：

"方案显示建筑的檐口带有

四段朴素简洁的带饰线脚。鸿达用朴实而现代的外立面替换之前的拱形底座，并采用裁成方形的石材饰面。透视图中还有一部象征现代生活的汽车。这一细节体现了鸿达将摩登建筑置于日常生活环境的良苦用心。"

鸿达的摩登方案很美好，但现实很不幸。这张美丽的效果图刊登仅3个月后，大楼着了火，顶楼被烧毁，其他楼层遭到严重破坏，损失大约有50万两白银。

1941年太平洋战争爆发后，日本人接管惠罗公司，将英籍经理关入集中营。1945年抗战胜利后，公司虽恢复营业，但因国民政府进口关税高，加上来自南京路其他百货公司的竞争，难以再现辉煌。1955年，惠罗公司歇业，由国营中百公司接收。

不知为何，1986年惠罗公司大楼不幸再次失火，建筑内部损毁严重，由鸿达设计的装饰艺术风格的建筑细节不幸付之一炬。现在的

惠罗大楼淡米色立面简洁平淡，历史细节不多，应该与大火有关。

如今，惠罗百货保留了百年前的名字，主营羽绒和丝绸，做南京路的游客生意。由于顾客老龄化加上电商竞争，这家国营老店引入设计时尚的苏州绣娘品牌，同时也进行了修缮，希望为这座老建筑引入新的活力。

昨天：惠罗百货　**今天**：惠罗商厦　**地址**：南京东路 100 号　**建于** 1906 年（1930 年改造）
设计师：鸿达（1930 年改造）　**建筑风格**：装饰艺术风格

The Whiteaway Laidlaw & Co Ltd building still retains its old Chinese name "Huiluo", but it's hard to tell the store's rich history from its current appearance. The department store was unfortunately struck by fire twice in 1930 and in 1986.

Established in Calcutta, India, in 1882, Whiteaway Laidlaw & Co. Ltd. was at one time the largest foreign-owned department store in modern Shanghai.

As the originator of Shanghai retail shops, Whiteaway Laidlaw & Co. Ltd. was an exquisite classic building that stood on a prominent corner of Nanjing and Sichuan roads in 1906. It was renovated into a chic Art Deco build-ing in 1930 with a beautiful arcade, curves and geological lines.

The store's slogan written in Art Deco style was "Quality, Value, Service" and began dealing in merchandise and importing business from 1904. The five-floor building came up in 1906. At first, only two floors were for shops while the upper three floors were leased out to foreign enterprises and clubs. As business grew, they took back the upper floors in 1919 for more space.

In the 1920s, Whiteaway Laidlaw & Co. Ltd. jostled with Lane Crawford & Co., Hall & Holtz Ltd. and Weeks & Co. Ltd. as the four largest foreign-owned department stores in Shanghai. The company enjoyed a "golden era" in the 1920s and 1930s when British pound was devaluated to silver, the widely used currency in China. The company took advantage and started importing commodities from London and sold them in Shanghai for silver, which they exchanged for pounds to import European products and built its fortune.

When Mence Wilkinson, chairman of Whiteaway Laidlaw & Co. Ltd., visited Shanghai in 1921, the *Shanghai Times* called him a "great business magnate".

"There is only one way of creating a great business of the type of Messrs Whiteaway and Laidlaw, and that is to deal fairly and generously with the public," the visiting chairman's opinion was reported in the 1921 article.

"One day we received a letter from a planter up-country ordering a dozen pairs of socks," Wilkinson was quoted in the interview. "In the letter the man hoped that the socks would be of better quality than those which had been sent to him in response to his last order. So we sent him a dozen pairs for nothing," said the chairman, who also gave other examples of the way Whiteaway Laidlaw did business.

"Generosity on this scale is bound to be rewarded, and the present status of the firm is testimony to the wise and thorough way in which the policy has been carried out by Wilkinson, his co-directors and his carefully selected managers. Time is not deemed to be wasted when it is employed in the task of completely satisfying the great public that is served by the undertaking," the report claimed.

The success Whiteaway Laidlaw & Co. Ltd. enjoyed led to the big reconstruction in 1930, "which will bring the premises of this well-known emporium into line with the most up-to-date buildings of Europe."

"For the purpose of obtaining a maximum of daylight and natural ventilation, the entire outside of the building will be changed, the masonry relieved from the out-of-date ornamentation and nearly the entire wall space opened up and provided with modern steel casement windows," said a report in *North-China Herald* in 1930.

"In harmony with this structural alteration, the refacing of the building will be executed along the modern lines of the architectural style which now has been generally adopted abroad. Sim-

plicity in design, dignity and effectiveness will be the keynote of this remodeled building."

The report says the architect was C. H. Gonda, a Hungarian architect whose works include the Cathay Theater on Huaihai Road, No. 14 on the Bund and Capitol Theater, all in ultra-modern style.

However, only three months after the report, a big fire gutted the top floor of the building and badly damaged the floor beneath, causing damage at Tls. 500,000.

The Japanese took over the building in 1941 and sent the British managers into Japanese camps.

Whiteaway Laidlaw & Co. Ltd. resumed business in 1945, but the "golden era" never returned after the Kuomintang regime slapped a steep import tax. In 1955, it was taken over by Chinese government and is now home to the state-owned Shanghai Huiluo Co. Ltd..

For a long time Huiluo was famous for hosting an exposition themed in down jacket from September to March every year. The store's silk products were also popular among tourists of Nanjing Road. But disaster struck again when another big fire destroyed much of the Art Deco building in 1986.

Facing fierce competition from online shopping in the 21th century, the store underwent a restoration and invited a new brand, Xiuniang Silk from Suzhou. Hopefully it will bring younger customers and new life to the time-honored department store.

Yesterday: Whiteaway Laidlaw & Co. Ltd.   **Today:** Huiluo Co Ltd
Address: 100 Nanjing Rd. E.   **Built in:** 1906 (renovated in 1930)
Architect: C. H. Gonda (for the 1930 renovation)   **Architectural style:** Art Deco

惠罗公司大火

Spectacular Fire at the Whtieaway Building

11月4日上午，一场大火完全烧毁了南京路四川路拐角处的惠罗公司顶楼，也严重损坏了其他楼层。同时，惠罗公司的营业场地都浸泡在水里。火灾造成的损失估计有50万两白银。

面对这项一段时间来最艰巨的任务之一，上海消防队表现出色，在接到火警警报后只用了不到一小时一刻钟的时间，就控制住了火势。

由于消防队里几乎每个人都在与火焰搏斗，南京路的早间交通因此被封锁了相当长的时间。为了排除拥堵，警察不得不把车辆交通引导到人行道上，因为南京路很快就堵车了。

今天早上一共有18台消防机器用于灭火，消防队的设备包括3个救生梯、水龙带和灭火龙等，从南京路码头泵取黄浦江水来灭火。

最早发出火警消息是一名巡捕，然后一分钟之内，位于沙逊大厦（今天的费尔蒙特和平饭店）中的火警瞭望塔也报告发现烟情。由于大楼重建工程已经进行了一段时间，整个惠罗大楼都被竹篱笆包裹着，屋顶覆盖着竹席屋顶，这使得对建筑物的观察变得非常困难。

很明显，火灾发生在用作储藏室的五楼阁楼，并从这里蔓延至屋顶和上方的席子，又烧到四楼。三楼只有一处着火点，消防员迅速检查了这里。

由于大楼内既没有消火栓也没有可连接的水泵，消防员必须将水龙带一直铺设到楼顶。第一条水龙带是沿着四川路入口的楼梯铺设，消防员铺设水龙带的巧妙工作对于灭火非常重要。

实践证明，使用救生梯将水龙带连接到较高楼层非常困难。当救生梯顶部碰到竹篱笆时，它仍然距离建筑物本身至少六英尺。在离地40英尺的高度，消防员设法拆下竹篱笆，并越过以到达楼层。两名消防员因掉落的竹子而受轻伤。

位于四层的建筑木板和支架极大地阻碍了消防工作，楼梯表面也同样阻碍了水龙带的移动。因为当时正准备重新铺上马赛克砖，楼梯表面被打磨成非常粗糙的状态。

整个惠罗商店里面都被水浸湿了，必须在每层地板上打孔以帮助排水。消防队的每张防水纸和上海其他地方能马上借到的防水纸，都被用来遮盖商店里的货物。

在消防队出动几分钟后，警察就在绕行的南京路和四川路忙碌工

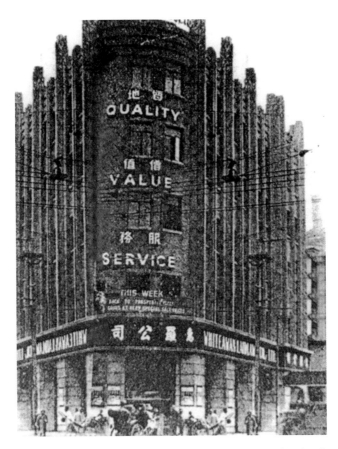

作。消防车的水带覆盖了外滩路面，因此人们必须绕行去上班。在一个案例中，大连丸号（Dairen Maru）的乘客因此无法在启航前及时赶到，被留了下来。大约到了中午，交通恢复正常。

Fire on the morning of November 4 completely gutted the top floor of the Whiteway Laidlaw building at the corner of Nanking and Szechuen Roads and badly damaged the floor beneath while many of the sales-rooms of the concern were soaked with water. Damage done is estimated roughly at Tls. 500,000.

Facing one of the most difficult tasks that has been presented to them for some time, the Shanghai Fire Brigade did good work and had the blaze under control in less than an hour-and-a-quarter from the time to alarm was given.

Morning traffic in Nanking Road was blocked for a considerable time as nearly every unit in the Brigade battled with the flames. To clear the block, the police had to divert traffic down side-streets as in a short time Nanking Road was jammed.

Eighteen machines turned out to this morning's fire, the Fire Brigate equipment including three turn-table escapes, while both floats, the Fire Dragon and the Mih ho loong pumped water from the Whangpoo at Nanking Road Jetty.

The first alarm was given by a police constable, and within one minute the Fire Brigade watchtower in the Sassoon building (today's Fairmont Peace Hotel) reported smoke. The whole Whiteaway building is at present encircled with a bamboo fence and the roof covered with a bamboo matting roof during the reconstruction work which has been in progress for some time past, and this rendered observation of the building very difficult.

The fire apparently broke out in the attic on the fifth floor, used as a store-room, spread to the roof and the matting above, and then to the fourth floor. Only in one place did it come through to the third floor, being quickly checked by the firemen. The roof of the adjacent Laidlaw building in Szechuen Road was also destroyed.

With no internal hydrants nor pumping connections in the building, it was necessary for the firemen to lay hose all the way up to the top of the building. The first line was taken up the stairway at the Szecuen Road entrance, and the smart work carried out by the firemen here was responsible to no small degree in holding the fire.

Work with the escapes to take lines of hose to the upper floors proved extremely difficult. When the top of an escape touched the bamboo fence, it was still at least six feet from the building itself, and firemen who managed to tear down sections of the fence and then cross to the stonework did so over a drop of 40 feet. Two firemen were slightly injured by falling bamboos.

Construction planks and trestles on the fourth floor hampered the

RECONSTRUCTED PREMISES OF
WHITEAWAY, LAIDLAW & Co. Ltd.
SHANGHAI

work of the brigade greatly, as did the surface of the staircase up which hose was taken. This latter has all been broken up preparatory to being refaced with mosaic tiles, and was in a very rough state.

The whole of the interior of the Whiteaway Laidlaw store was sodden with water, and holes had to be made in every floor to assist to dispersal of the water. Every water-proof sheet of the Brigade was used in covering up goods in the store, as well as every other sheet in Shanghai that could be borrowed at short notice.

Within a few minutes of the Brigade turning out the police were busily at work in detouring traffic Nanking Road, Szechuen Road and the Bund were covered with fire hose the latter from the floats, and as a result Shanghai went to work by varied and tortuous routes. In one case passengers by the Dairen Maru were unable to get to the ship before it sailed and were left behind. Traffic was back to normal about noon.

摘自 1930 年 11 月 11 日的《北华捷报》
Excerpt from *The North-China Herald,* on November 11, 1930

灰色大楼的繁华与忧伤

A Grey, Eventful Building

灰色慈安里大楼古典庄重，但从白色券窗里伸出一根根晾衣杆，仿佛想要透露一言难尽的故事与沧桑。

一个多世纪前，这座砖木结构的建筑是犹太富商哈同投资的商务楼。高达5层的大楼呈L型，建筑面积约为4709平方米，由英商爱尔德洋行设计为安妮女王风格。大楼底部两层是银行、钱庄、商铺和库房，三四层用于租赁办公，顶层是出租的住宅。

在大楼建成的1906年，这种新古典主义的安妮女王风格正在风行。同济大学常青院士研究发现，19世纪70年代后，南京路外滩段的建筑从外廊式风格向安妮女王复兴风格演变。这种新古典主义建筑风格由英国建筑师理查德·诺曼·肖（Richard Norman Shaw 1831–1912）开创，以清水红砖墙面为特征，饰以白色面砖、线脚和其他装饰元素，追求一种富贵华丽的视觉效果。

慈安里大楼的灰色清水砖立面，装饰有白色阳台和连续的白色券柱式窗套，灰白分明。常青

院士提到，大楼后期加建的老虎窗具有巴洛克风格，因此也有人把它归入折衷主义风格。南京路汇中饭店和江西中路礼和洋行大楼也是这种风格的建筑，都是清水红砖，离慈安里大楼仅几分钟之遥。

灰色大楼是哈同众多的南京路项目之一。哈同出身贫寒但头脑灵活，他从沙逊洋行的看门人做起，后来担任房地产业务主管。他转入新沙逊洋行工作后眼光独到，认为南京路独特的枢纽位置"其繁盛必为沪滨冠"，将投资的重点放在南京路，低价购

入南京路的地产。南京路两侧以"慈"字命名的大楼、里弄大多曾是哈同的产业。而南京路果然如哈同所预见的成为一条最繁华的中华商业街，堪称上海版的"纽约第五大道"。

慈安里大楼的命运与南京路紧密相连。大楼随着南京路的发展而兴盛。1908年，上海第一辆有轨电车从大楼脚下经过，从外滩驶入南京路，1930年代大楼加建了巴洛克风格的山花和老虎窗。抗战期间上海租界成为"孤岛"，许多避难的人占据了空置的大楼，灰色大楼功能逐渐向居住改变，直至今日。昔日大房间被分隔成为多个小隔间，拥挤杂乱。如今，大楼里居住着150多户人家，每户只有十几个平方米，以老人和租客为主。

热闹繁华的南京路，每天不知有多少熙熙攘攘的游客走过灰色大楼，却很少人停下脚步仔细看看大楼的一块勒石，上面刻着

"茅丽瑛烈士遇害处"。这块不起眼的勒石为灰色大楼平添一丝忧伤的气息。

茅丽瑛是一位出身贫寒，但乐观上进的杭州才女。历史照片上的她短发旗袍，笑容温暖。她毕业于启秀女中（今上海市启秀实验中学），曾在上海海关工作。1938年5月5日"上海中国职业妇女俱乐部"在慈安里大楼成立，茅丽瑛当选主席。

1939年，为支援新四军抗日斗争，她克服日伪特务百般阻挠，在位于慈安里大楼二楼的福利公司义卖募集寒衣。12月12日晚，她步出大楼时遭到特务枪击，三日后去世，年仅29岁。

1989年，上海为茅丽瑛举行纪念活动，在慈安里大楼遇害处勒石纪念，启秀实验中学也设立了一尊她的雕像。校园中的茅丽瑛雕像短发旗袍，浅浅的微笑坚定温暖，仿佛还在灰色大楼里忙碌着她热爱的事业。

昨天: 慈安里大楼　**今天**: 慈安里大楼　**地址**: 南京东路 114-142 号
设计师: 英商爱尔德洋行　**建筑风格**: 安妮女王复兴风格
参观指南: 建议参观同为安妮女王风格的江西中路 255 号礼和洋行。

Ci'anli Building is a five-story L-shaped building in Queen Anne style. When it was built in 1906, the building had shops, banks and storage areas on the first two floors, large offices on the second and third floors, and rental flats on the top.

According to the research of Tongji University Professor Chang Qing, the neo-classical Queen Anne style was in vogue when the building was constructed. "After the Second Opium War (1856-1860), Western expatriates grew more confident about their future in China, so they began construction on a larger scale. From the 1870s to the 1920s, an earlier colonial style was gradually replaced by more authentic Western architectural styles. With the spread of external brick walls, facades with colonial verandas fell out of fashion and were replaced by arcades, pediments and brick-carved patterns. It showed the influence of the Queen Anne style," the professor notes in his book *Origins of a Metropolis: A Study on the Bund Section of Nanjing Road in Shanghai.*

Architect Richard Norman Shaw (1831-1912) popularized the Queen Anne style of British architecture, beginning in the industrial age of the 1870s. It was char-acterized by external brick walls, white tiles or architraves and elaborate motifs reminiscent of Baroque ornamentation. Almost at the same time, buildings of this style began appearing in Shanghai. The window pockets of the Ci'anli Building were decorated with red brick frames or white arcades. Dormer windows jutted out from below a pitched roof.

Professor Chang listed several examples of this style remaining in the Nanjing Road neighborhood, including the Carlowitz & Co. Building on Jiangxi Road M. and the building on 199 Dianchi Road.

The Ci'anli Building was part of grand-scale Nanjing Road development initiated by Jewish tycoon Silas Aaron Hardoon. He was born in Baghdad and educated at a charity school in Mumbai after his family moved to India.

In 1868, Hardoon arrived penniless in Shanghai, where he was employed by David Sassoon & Co. as a rent collector and watchman. His genius for commercial property acquisition led him to become the largest single owner of real estate in the city's international settlement. While working for the Sassoon company, he decided to focus on investment along Nan-

jing Road after carefully analyzing the development of Shanghai real estate.

Hardoon believed that Nanjing Road would be the most prosperous place in Shanghai due to its location. So he purchased numerous properties along Nanjing Road for the Sassoon family and for himself, using his savings and rents he collected. The value of these properties soared as Nanjing Road grew ever more prosperous, just as he had foreseen.

Hardoon subsequently founded his own company and made huge profits from investments on Nanjing Road, which came to be called "Shanghai's Fifth Avenue". He served on the municipal councils of both the French concession and the international settlement.

The layout of the Ci'anli Building, which bears neither a grand vestibule nor an eye-catching entrance, was designed purely for functionality. That exemplified Hardoon's approach to maximizing profits from his real estate investment.

When Hardoon died in 1931, his estate accounted for about 44 percent of the total properties on Nanjing Road. Since then Hardoon company was managed by his adopted son George but many of his holdings were occupied by Japanese soldiers during World War II.

On December 12, 1939, Mao Liying, a Communist Party member and chairwoman of the Shanghai Professional Women's Club, was assassinated at the gateway of the Ci'anli Building. She was organizing a charity bazaar to raise winter clothes for New Fourth Army soldiers who were fighting with the invading Japanese military force. A stone plaque on the facade today commemorates this young, short-haired girl from Hangzhou.

Yesterday: Ci'anli Building **Today:** Ci'anli Building **Address:** 114 Nanjing Rd. E.

Built in:1906 **Architect:** Algar & Co. **Architectural style:** Queen Anne

Tips: The building is under renovation, so those seeking to view a structure of similar architecture are directed to the Carlowitz & Co. Building a few minutes' walk away at 255 Jiangxi Road M..

礼和洋行大楼
The Carlowitz Building

江西路255号礼和洋行的建筑风格与慈安里大相似，接近安妮女王风格，但其外墙是用红砖而不是灰色砖覆盖的。

这座砖木结构的建筑建造于1898年，不久就被德商礼和洋行（Carlowitz&Co）收购作为中国总部使用。礼和洋行的业务是代理德国重型机械、采矿设备、军火和精密仪器等工业品，后来发展成为远东地区最大的德商贸易洋行之一。洋行为总部大楼的选址位于上海市中心，距离"上海第五大道"南京路只有几步之遥。

礼和洋行大楼是19世纪后期上海公共租界体量最大的西方建筑之一。红砖墙面点缀有白色的柱头、线脚、宝瓶栏杆等元素，装饰精美。更引人注目的是美丽的外廊，底层为连续半圆拱券，其他楼层是连续平弧形券柱廊。大楼顶部是巴洛克式的山墙。

1949年以后，建筑由黄浦区政府的多家单位使用。这座美丽的建筑近年历经修缮，恢复了曾被封闭为室内空间的外廊，成为一个商业办公综合体。

The Carlowitz Building on 255 Jiangxi Road M. mirrors the architectural history of the nearby Ci'anli Building, but its facade is red brick, not gray. The brick-and-wood structure was erected in 1898 and purchased by German firm Carlowitz & Co. as its China headquarters. The company sold machinery and was one of the largest German trading companies in the Far East at the time.

The company's importance was emphasized by the location of its offices only a few steps from Nanjing Road - then called "Shanghai's Fifth Avenue". The enormous structure was among the largest Western-style buildings in the area.

The redbrick facade of the building exquisitely displays all-white architectural ornamentation. Even more eye-catching are the verandas with semicircular arcades on the ground floor and smaller arcades on

the floors above. Baroque-style gables top the building.

After the founding of the People's Republic of China in 1949, the building housed offices and institutions of the Huangpu District government.

After a restoration, the building regained the beauty of historical look and was transformed into a complex of shops and offices.

梦幻拱廊的前世今生

1930s Central Arcade Gets a New Makeover

1930年的上海，位于南京路黄金地段的中央拱廊落成，璀璨明亮，引人注目。

中央拱廊的开发商是犹太商人爱德华·埃兹拉（Edward Ezra）。埃兹拉家族与沙逊等塞法迪犹商家族一样，长期投资经营上海的房地产项目，实力雄厚。

这个商业地产项目由四条马路——今日的南京路、九江路、四川中路和江西中路围合而成，由四座兴建于1924到1930年间的建筑组成，分别是中央大楼、美伦大楼、新康大楼和华侨大楼。徐家汇藏书楼馆藏的犹太期刊《以色列信使报》详细介绍了这个项目。

"美丽明亮的拱廊与四川路和九江路平行，占地5亩，店铺琳琅满目。基地被宽阔的爱德华·埃兹拉路和四川路分成四部分，南京路和九江路之间的部分建有天顶。中央拱廊的底层是上海股票交易所和带中央供暖设施的摩登店铺，楼上为办公和公寓。"1930年2月7日的《以色列信使报》介绍。

文章还提到，古典风格的建筑由公和洋行设计，比例优美，立面设计还考虑到与南京路江西路转角的美伦大楼保持和谐的关系。

同济大学钱宗灏教授认为，该项目是上海最早的拱廊建筑之一。法国大革命后，人们在既有建筑相对称的小街道上搭起用铸铁和玻璃构筑的天顶，既透光，又隔绝灰尘和恶劣天气，深受新兴中产阶级的喜爱。至今犹存的巴黎维维安拱廊就是一例。拱廊里遍布富有人文气息的商店和温馨的咖啡馆，适宜漫步和休闲。

在19世纪的欧洲，拱廊建筑一度盛行，但后来受到大型百货公司的竞争，渐渐地不再流行。而随着时代发展，上海南京路的中央拱廊也经历了戏剧化的功能变迁。

1945年二战结束，中央拱廊聚集了不少流动商贩，出售美军剩余物资和走私洋货，这里因此被称为中央商场。美国著名导演斯皮尔伯格以上海为背景的电影《太阳帝国》里也有关于中央商场的镜头。

1949年后中央商场公私合营，以售卖残次不合格品为特色。因为价格实惠又无需凭票购买，这些有瑕疵但不影响使用的商品在那个物质匮乏的年代十分畅销。商场还提供修理服务，能修理好别处修不好的物品，受到大众青睐，周日常人山人海。

2005年，黄浦区对南京路东段进行商业改造，中央商场关门停业，于2017年一期改造竣工后对外开放，变身名为"外滩·中央"的商业地产，开出很多时尚店铺。著名的纽约林肯爵士乐中心也在美伦大楼设立了剧场长期驻演。昔日的梦幻拱廊历经好一番变迁，渐渐回归初建时的定位。

1929年的秋天，中央拱廊项目竣工在即，英文《大陆报》报道透露，项目由名为索普（Messrs. Arthur and Thedoor Sopher）的外侨负责运营，将开设很多精美店铺，其中包括马克思·歌瑞尔（Max Grill）以新奇商品著称的商店。歌瑞尔在青岛开设了很大规模的百货商店。

"他正在上海，为南京路55号中央拱廊开幕做准备。在这家新店里，他将销售自己近期专程赴欧洲采购的最时髦的商品。"《大陆报》报道写道。

报道还提到，歌瑞尔有丰富的零售业经营经验，他能够保证上海店的货品不会流于平凡，时髦的新店可以与欧洲大都市的百货商店媲美。

同济大学郑时龄院士介绍，商场的拱廊街原来可以从南京路通到九江路，现已不存。也许为了尽量重现1930年那璀璨明亮的拱廊，"外滩·中央"项目的楼内还有一条内廊，设计效果图显示，未来计划在十字街上方加建玻璃天顶。

昨天: 中央拱廊　**今天**: 外滩·中央　**地址**: 南京东路 119-137 号　**建于**: 1930 年
设计师: 公和洋行　**建筑风格**: 新古典主义
参观指南: 请注意中央大楼与相邻美伦大楼立面风格的和谐之妙。

A particularly well-lit arcade of the Edward Ezra Arcade at Nanjing and Jiujiang crossroads was widely covered by Shanghai's English newspapers in 1930.

Part of that estate was reopened to the public after a redevelopment project titled "The Central". A huge glass roof which attempts to bring back the 1930s look will cover four historical buildings in the commercial block — Central Mansion, Meilun Building, Xinkang Building and Huaqiao Building — all built between 1924 and 1930.

"Encircled by Sichuan, Jiangxi, Jiujiang and Nanjing roads, this commercial block was one of the city's earliest shopping arcades that was influenced by the early European classical arcade architecture such as Galerie Vivienne in Paris," says Tongji University professor Qian Zonghao, author of the book *Shanghai Nanjing Road*.

The Edward Ezra Arcade, later called the Central Arcade, was designed with an iron frame and covered by a glass roof. Arcade architecture was a popular social center for the "new bourgeois class" which emerged in the early 19th century after the French Revolution and required urban public space for shops, cafes, salons and squares.

"The pedestrian arcade and crisscrossing blocks offered an attractive place for shops and pleasant lighting where people could stroll, browse and linger," Qian says.

The estate owned by Jewish tycoon Edward Ezra was introduced in the city's major Jewish publication *The Israel's Messenger* in 1929 and in 1930.

"A handsome and well-lit arcade with shops on either side runs (that) parallel to the Szechuen and Kiukiang roads, covering an area of 5 mow. The site is divided up into four sections by the wide private Edward Ezra and Szechuen roads forming a covered way between Nanking and Kiukiang roads. The whole of the Ground Floor, with the exception of the Shanghai Stock Exchange premises, is occupied by modern shops with central heating," a report in The Israel's Messenger on February 7, 1930, said.

The report also noted that the exterior was designed by Palmer & Turner in a classical manner with well-proportioned openings, and the facade was designed keeping in mind the existing structure on the corner of Nanjing and Jiangxi roads. That structure was

Meilun Building that also came up in 1930.

However, the popularity of shopping arcades in Europe declined with competition from large department stores or urban redevelopment.

After World War II which ended in 1945, a group of small vendors began congregating here selling pens, lighters, second-hand watches, clothes, etc. Soon after a large quantity of surplus American military goods was dumped here which gradually became a popular market known as the Central Market. With 90 counters at pricey rent in the late 1940s, the market was also seen in Hollywood director Steven Spielberg's movie "Empire of the Sun".

An interesting bit about the shopping arcade came to light after the 1960s. A document at the Shanghai Archives Bureau claimed that the market later started selling "unusual, novel, rare and superior second-hand products ..."

"Products that were hard to buy elsewhere were often available in this market, which had 130 shopping counters in seven departments including clothes, medical appliances, chinaware and repair," the archive report says.

It became an attractive market for defective goods in such a central location. The price was only half or even one-third of the market price, such as "nude batteries," or batteries with adequate power but no wraps. Many locals found it economical to buy components one by one and then assembled a radio, an electric fan or even a bicycle. The market could also fix "unrepairable stuff," from plastic slippers, radios to table tennis bats used by Chinese world champions.

As European shopping arcades had seen a revival over the past decades, this majestic architectural complex closed in 2005 for an overhaul which was done by Shanghai Bund Investment (Group) Co. Ltd..

After the project completed around 2021, The Central will be a commercial landmark for upmarket shops and fine dining. The New York-based Jazz at Lincoln Center orchestra had opened a 500-square-meter Shanghai branch in the Meilun Building.

Zheng Shiling, a scholar on architecture and one of the experts supervising the project, says a glass roof will cover the cross streets of the project. While it may not be possible to restore its 1930s

look, a corridor will link the historical buildings.

When the Central Arcade was "rapidly nearing completion" in the autumn of 1929, the China Press reported that there were many fine shops including a novelty store of Max Grill, who was the owner of the largest department store in Tsingtao.

"He is at present in Shanghai making arrangements to open on the first day of December a fancy goods and novelty store at 55 Nanking Road, Central Arcade, where he will stock all the latest novelties from Europe which he specially secured during his recent visit there," the report said.

The story also noted that Grill was an old hand in the store business who "promises the Shanghai public something out of the ordinary, a novelty store that will be equal to those in the capitals of Europe, replete with all the latest hits."

"History does not repeat itself, but it often rhymes" is a quote often attributed to Mark Twain. The architectural history of Shanghai seems to confirm that.

Yesterday: The Central Arcade   **Today:** The Central   **Address:** 119-137 Nanjing Rd E.

Built in: 1930   **Architectural style:** Neo-classical

Tips: Please note the similarity in architectural style between the Central Mansion and the adjacent Meilun Building.

中央拱廊开业的第一家店
The Carlowitz Building

新大楼中央拱廊位于四川路77号靠九江路的转角处，这里开业的第一家店是山田（J. Yamada）的美发沙龙和理发店，他们位于南京路18号B的沙龙一直营业到最近。

由于拆除了四川路临时的木板屏障，公众首次看到了新建筑的迷人外观，其面向街道和拱廊的商店布置巧妙，从九江路和四川路直通爱德华·埃兹拉路，以后还将延伸到南京路。

虽然只有三张理发椅子，山田美发店因为地段方便的优势，在不愿浪费时间的商务人士间非常流行。山田先生希望在三四天内为店铺配备齐全的设备，可以为顾客提供服务。山田美发店的两面都临街道，很有优势，山田先生将面向四川路的那一侧用于服务理发客人，而面对拱廊的一边则留给女士们使用。一位外籍女士将负责这部分的女宾服务，她在洗发、烫发、美甲和持久电发方面拥有丰富的经验。

The first shop open for business in the new building known as the Central Arcade, at 77 Szechuen Road, at the corner of Kuikiang Road, is the hairdressing Salona and Barber Shop of J. Yamada, who until recently operated the Valet Salon at 18-B Nanking Road.

With the removal of the temporary board screen on Szechuen Road, the public got its first glimpse of the attractive appearance of the new building with its clever arrangement of shops facing the streets and arcade, which leads from Kuikiang and Szechuen Roads straight through to Edward Ezra Road and which later will extend to Nanking Road.

The shop of Yamada began its popularity immediately with the patronage of businessmen who lost no time in taking advantage of the opportunity of conveniently located service, even though there were but three chairs installed and operating. Within three or four days Mr Yamada expects to have his shop entirely equipped and in complete working order for his patrons. Having the advantage of two frontages, Mr Yama-

da is devoting that part of his shop facing Szechuen Road to the patronage desiring tonsorial service, while facing the arcade is a section reserved for ladies.

A foreign lady will be in charge of this part of his shop. She has had much experience in shampooing, marcelling, manicuring and permanent waving.

摘自 1929 年 2 月 3 日《大陆报》
Excerpt from *the China Press*, on February 3, 1929

充满力量的电力大楼

A Powerful Building for a Power Company

1931年，美商上海电力公司位于南京路的新厦落成。大楼高达6层，立面强调竖向线条，檐部有精美的细部处理，是一座典型的装饰艺术风格建筑。

上海电力公司（Shanghai Power Company）1882年由英国人创办，是上海最主要的电力供应商，1928年后由美商经营，一直营业到1949年。

位于南京路181号的新大楼在底层设有电器展示厅，向消费者宣传用电的好处。大厅用大理石装饰，陈列着琳琅满目的家用电器，如烤炉、热水壶和取暖器，客户既能购买也可以租赁。在搬到南京路181号前，公司的展示厅设在南京路靠近外滩的沙逊大厦（今费尔蒙特和平饭店）。

当时，国人的观念传统保守，对于"电"这种西人的发明缺乏了解。为了拓展市场，上海电力公司发挥创意，用雷公电母等中国神话人物来介绍电能。值新大楼落成之际，公司又利用媒体宣传电能的优点——清洁无烟、使用方便、价格适中。

"只需一般花费，就能拥有该公司租售的电器，其带来的舒适超过一个世纪前帝王的享受。在天冷的季节，起床能看到电热器那耀眼的光芒，用电加热的水洗个热水澡，然后享用由电炉烹制的早餐。" 1931年12月21日《大陆报》的报道写道。

电力公司的新大楼是一座坚固的钢筋混凝土建筑，虽然仅有6层楼高，但其坚实的结构允许未来有需要再加盖三层。

大楼由美国建筑师哈沙德设计，他是近代上海最重要的建筑师之一。1879年，哈沙德出生于南卡罗来纳州一个大米种植园家庭，他曾在格鲁吉亚技术学院学习建筑，毕业后在纽约知名事务所任职，1905年在第五大道开设自己的事务所。他在纽约的建筑实践为日后在上海的设计成就打下伏笔。

1921年他到上海发展，先与另一位美国知名建筑师——茂飞（Henry Murphy）合作。1923年茂飞关闭上海办事处后，哈沙德留了下来，把当时风靡美国的布扎美术风格和意大利文艺复兴

风格等移植到了上海的建筑作品上。他后来转向摩登的装饰艺术风格（Art Deco）。这种以几何、放射状图案为特征，有强烈装饰意图的风格源于1925年法国巴黎博览会，很快在包括上海在内的许多大都市流行。

电力公司大楼是哈沙德在上海的晚期作品。负责修缮大楼的建筑师沈晓明发现，这座美商电力公司与同时代美国纽约建筑有惊人的相似之处，如壁柱的处理、转角装饰和几何纹样等。

回到摩登的1931年，Art Deco这种象征光明、活力和力量的建筑风格，也许是一座电力公司大楼最好的选择。大楼外立面饰有垂直线条和泰山面砖，顶部有凸出的白色装饰，窗户下面有青铜板装饰，与墙面颜色形成对比。除了充满力量的外立面，大楼室内也有不少电力主题的细节，如设计为闪电状的铸铁栏杆。

大楼建成后，一楼为展示

厅，二到五层是办公室，顶楼是用柚木考究装修的套间，供公司高管使用。

如今，这座历史建筑与相邻的建于20世纪80年代的华东电管大楼一起，被改造为低调奢华的艾迪逊酒店（Edition Hotel）。

沈晓明认为，从此南京东路上又多了一座市民可以走进去的历史建筑。

"南京路这么重要的位置，多开放一幢建筑，魅力就多一点，要让历史的街道和建筑融入到现代城市生活中去，"他说。

昨天：美商上海电力公司　**今天**：艾迪逊酒店（The Shanghai Edition Hotel）
地址：南京东路181号　**建筑时间**：1931年　**建筑师**：哈沙德（Elliot Hazzard）
建筑风格：装饰艺术风格 Art Deco　**参观指南**：大楼作为艾迪逊酒店使用，电力主题的历史细节保存好。

The power company moved into the building at 181 Nanjing Road in 1931 and invited people to watch the demonstration of electrical apparatus because "everything electrical is here — for rent or for sale".

Shanghai Power Company was the city's main electricity supplier until 1949. The company used the images of Leigong, God of Thunder, in Chinese mythology, in its advertisements to promote the use of electric power among the locals.

The company also used the media to explain the advantages of using clean power with no dirt or smoke, affordability and convenience of use.

To make its case, the ground floor of the building boasted an elaborate marble-adorned showroom to demonstrate electrical apparatus including cooking stoves, hot water heaters and radiators. The showroom was previously at the Sassoon House also on Nanjing Road.

"With what the company rents and small purchases as may be desired — like electric percolators and electric fans — comfort beyond the dream of kings a century ago may be had at moderate cost. One may arise in the cold-est weather to the vivid glow of an electric radiator, bathe in water electrically heated, and eat breakfast cooked on an electric stove," a report in *the China Press* on December 21, 1931, said.

The building is a reinforced concrete and steel structure "sturdy enough to withstand all the blows to which buildings are susceptible for many years." Though it had only six stories, the structural strength of the building allowed for the addition of three more stories if necessary.

American architect Elliot Hazzard used the most modern technology that was in vogue at that time such as the automatically controlled Ottis elevators — the fastest in the city in 1931. Hazzard also designed three other signature buildings on Nanjing Road — Wing On Tower, Foreign YMCA Building and China United Apartments.

Hazzard was an influential architect in Shanghai in the late 1920s and early 1930s. The early period of his architectural career was dominated by eclectic style but had switched to Art Deco style in the later years. The Shanghai Power Company was one of the masterpieces of the late period.

Art Deco style, which took off after the 1925 Exposition des Arts Decoratifs in Paris, coincided with Shanghai's real estate boom from the 1920s to the 1930s and had a great impact on the city's architectural scene.

Art Deco buildings in Shanghai are generally divided into two styles — the French and the American Art Deco. The Shanghai Power Company building can be categorized into the "American Art Deco" as it mirrored Art Deco buildings in New York of the same period in many ways, such as its treatment of pilasters, decorations over the corners and geological patterns.

Back in 1931, Art Deco style that symbolizes brightness, vitality and power, might be a fitting choice for No. 181, which was owned and used entirely for the offices of Shanghai Power Company. Some surviving historical details reflect the history of the building, such as the patterns of lightning on the staircase railing or muscular men with a hammer or with lightening and clouds on the big iron gate.

In addition to the showroom on the ground floor, the second, third, fourth and fifth floors were occupied by staff members while the sixth floor was the executive suites. All the rooms on the sixth floor were paneled in teakwood and well-furnished. The dimen-

sion of each floor was about 150 by 100 feet.

The pleasing facade was decorated by Chinese Taishan tiles while cast-iron spandrels were used for the plate glass windows. The facade highlights vertical lines with the convex decorations in white color on the top. Below the window is the embossed metal board in bronze color that contrasted with the wall surface in colors.

The building has completed renovation and reopened as public area of the hotel, featuring restaurants, spa rooms and a convention center while the adjacent 1980s high-rise, East-China Electric Power Building at No. 201 housed the rooms.

The significance of this project, Shen Xiaoming, restorer of the building believes, is that it will open one more historical building to the public on Nanjing Road.

"Nanjing Road is such an important thoroughfare that the opening of one more historical building will certainly add more glamour to this street," he says.

Yesterday: Shanghai Power Company

Today: The Shanghai Edition Hotel (the part of public area)

Address: 181 Nanjing R.   **Built:** In 1931

Architect: Elliot Hazzard   **Architectural style:** Art Deco

Tips: Please admire Art Deco elements on the facade of the building. It's fun to explore 1930s power-themed decorations hidden inside the hotel.

上海电力公司展示各类冰箱产品
Shanghai Power's bid to boost fridge sales

上海电力公司的泰勒夫人说，再过几周上海夏季惯常的炎热就会到来了。她提醒那些尚未为暑热带来的不愉快做好准备的人，他们还未考虑到自己的健康与舒适。

五月是研究了解冰箱的月份，几乎每一个知名、可靠的电冰箱制造商都在本地市场有代表。上海电力公司位于南京路181号的展厅展示了各种型号和尺寸的产品。

上海电力公司家庭服务部的工作人员可以提供有关每种电冰箱的详细信息，并展示它们的各种优点。

公司认为没有必要提醒有想法的孩子父母，夏天比其他任何季节都应该注意妥善保存食品。在市场上精心选购的水果、蔬菜和肉类清洗后在电冰箱中冷藏保存，可以避免在潮湿空气里滋生细菌。变质的食物不易被察觉，往往要加热后或有了明显腐坏迹象才会被发现，但是几乎任何食物在超过50华氏度的温度存放都会变质。电冰箱可以确保有合适的温度、纯净的空气和冰块，因为电能本身就是纯净的。上海电力公司家庭服务部的泰勒夫人希望所有消费者和电冰箱的潜在客户都来参观南京路的陈列室，看看那里展示的富有吸引力的样机。

In a few mere weeks the hot dense heat of the usual Shanghai summer will be reminding those who haven't prepared for its various unpleasant contingencies that they've been thoughtless of health and comfort, says Mrs Taylor of the Shanghai Power Company.

May is the month for investigating refrigerators: almost every well-known and reliable manufacturer of electric refrigerator is represented on the local market. The S.P.C. has in their showroom at 181 Nanking Road various models and sizes from the representatives.

Men and women from the Home Service Department of the company can give detailed information on every kind of electric refrigerator and show the diversified advantages of each.

The S.P.C. feels there is no need to remind the thoughtful Shanghai parent that during this summer more than any other, the ultimate in food protection should be given. After careful selection at the markets: fruits, vegetables and meats, washed and placed in the pure cold temperature of the electric refrigerator are almost bound to be protected against the savages of humid germ filled air.

Spoiled foods cannot always be detected until after heat and dangerous bacteria has worked visible damage, but spoiled food is almost any food that has been left in a temperature above 50 Fahrenheit. The electric refrigerator guarantees a correct temperature plus pure air and pure ice because electric energy is in itself pure.

Mrs. Taylor of the Shanghai Power Company's Home Service Department hopes that all consumers and potential owners of electric refrigerators will visit the Nanking Road showrooms and see the attractive models on display.

摘自 1938 年 5 月 11 日《大陆报》
Excerpt from *the China Press*, on May 11, 1938

幸运的华电大楼

A Lucky Modern High-rise

在南京路上，华东电力调度大楼的历史不长，但故事很多。

1988年大楼建成时，它奇特的造型曾引发讨论与研究，后来荣获了建筑大奖。2013年大楼面临被改造为一间精品酒店，大幅改动立面的方案又引发争议，演化为一场"立面保卫战"，最终成为上海历史保护与城市更新的一枚经典案例。

华电大楼高达24层，是1949年后南京路建成的第一座高层建筑，1988年由同济大学毕业的两位建筑师——罗新扬和秦螯设计。那是一个高层建筑不多但思想已开始活跃的年代。

大楼形似一个巨大的立方体，21层开始有凸出的微波塔和斜屋面，延伸到顶部，在当时是新奇前卫的设计。两位建筑师后来谈到，微波塔、斜屋面和三角窗等设计都是从功能出发，与一些媒体加注的"后现代主义"标签关系不大。建筑师在设计时也考虑到建筑与外滩—南京路风貌区的关系，红砖立面与附近圣三一教堂的墙面十分协调。

华电大楼作为20世纪80年代建筑形式创新的重要案例，在业界屡获大奖，包括建国四十周年"上海十佳建筑"和建国五十周年"上海经典建筑"。但同济大学华霞虹教授透露，华电大楼建成后的一次调研显示，超过53%的公众认为设计很一般，6%的人甚至认为这座大楼很难看，与专业圈的高度评价形成鲜明对比。

上海出生长大的艺术家张兰生对这座大楼印象深刻，觉得设计有现代感，但是"在美学上不敢恭维"。

"这些年来，它所谓的后现代感似乎减弱了，整个城市众多的后现代主义、未来主义的建筑，使它不再招人眼球。"这位旅居澳洲的艺术家评价道。

2000年，华东电管局曾对大楼进行改造。电管局迁到浦东后，大楼由同属国家电网的鲁能集团接管。当时，黄浦区政府想对外滩—南京东路地段进行城市更新，提升风貌品质，带动区域旅游业发展。由于南京东路沿线精品酒店稀缺，鲁能集团希望将这幢办公楼和相邻的20世纪30

年代上海电力公司总部整合，打造成低调奢华的艾迪逊酒店，既满足南京路高档酒店的需求，也可以在这个黄金地段树立企业品牌。

2014年，鲁能集团邀请了四家中外知名建筑事务所，共设计了几十个方案送审。其中，大多数方案对红砖立面"改头换面"，有两个方案将立面改为装饰艺术风格的垂直线条。这些方案经同济大学陆地教授的微博转发后，引发沪上各界人士的大讨论。学者、媒体纷纷呼吁保留这栋建筑的外立面，留住城市共同的记忆。

其实，这些方案并非没有道理。因为要将两个不同时代的建筑融为一体，设计师将华电大楼的现代立面改为美商电力公司的装饰艺术风格，也是合理的方案。从法律上讲，在上海建成30年以上的建筑才会列入保护范围，1988年建成的华电大楼当时未到30年，并非一座保护建筑，立面是可以改动的。

上海市规划局为此专门组织专家会议，经过多轮讨论协调，要求保留约90%的外立面要素，原方案重新设计。

2016年4月28日，上海市建筑学会既有建筑更新研究设计中心还举办了一场有趣的沙龙活动，邀请政府、地产商、媒体社团和设计等四方代表，围绕"华东电调大厦改造案例"展开对话。

业主鲁能集团代表陈海涛坦言，项目的过程"非常费劲"，为了保住外立面，前后做了将近50版的方案，有种"曲线救楼"的感觉。负责改造设计的建筑师范佳山介绍，建筑的原有体量不能突破，外立面和历史要素——微波塔、坡屋顶、三角窗及外墙颜色都要保留，又要满足功能转换的要求，相当于"在一个桃子里塞进一只菠萝"。

"也许大家看到大楼立面成功地保住了，但未必了解其中的曲折，业主花费了大量的时间、精力和金钱。"她提到。

上海市建筑学会理事长曹嘉明曾参加规划局决定华电大楼命运的专家会议。他回忆，开始知道大楼要被"变脸"时很惊讶，

开会时专家们的看法高度一致，一定要保护。

"实现这个结果，各方都付出极大的努力，媒体发声，学会呼吁，规划局发文，业主从善如流，设计院无限配合。这个结局实现多方共赢。对于业主来说，华东电调大楼作为既有建筑改造的经典案例，提升其品牌的价值。对于生活在这个城市的居民来说，则是留住共同的城市记忆。"曹嘉明说。

在一次公众讲座上，华霞虹教授将华电大楼项目与巨鹿路888号案例进行对比。2017年，上海市优秀历史保护建筑——巨鹿路888号老洋房被业主在改造中私自拆除，原址建起一座金属混凝土结构的建筑。巨鹿路888号是国际饭店设计师邬达克于1920年代设计的22幢美式别墅之一，建筑破坏事件被曝光后，业主被罚款人民币3050万元，责令在10个月内恢复原状。

"这两个房子，南京东路201号和巨鹿路888号建造于不同时代，它们的风格很不同，但在改造过程中遇到了同样的问题。无论是巨鹿路888号被拆毁罚款，还是华电大楼的争议，都体现了我们城市各种力量对历史保护与城市更新的认识。"她说道。

昨天：华东电力调度大楼　**今天**：上海艾迪逊酒店　**地址**：南京东路 201 号

建造时间：1988 年　**建筑师**：罗新扬、秦壅　**建筑风格**：后现代

参观指南：建议欣赏大楼立面上历经周折得以保留的设计元素——大楼轮廓、三角窗、斜屋面、微波塔和建筑色彩。

The 24-floor East China Electrical Power Building was the first "super high-rise" to come up on Nanjing Road after 1949. Its modern appearance was weird to many people in 1988 and the attempt to turn the modern high-rise at 201 Nanjing Road E. into an Art Deco hotel in 2014 triggered a controversy that led to a drastic change of plans. The reworked plan had been hailed as one of the examples of urban regeneration in Shanghai.

The high-rise was designed during a dynamic "new era" — the late 1980s and early 1990s, just years after the reforms and opening up policy was taking effect in China. Two Chinese architects, Luo Xinyang and Qin Yong, both graduates of Tongji University, tried to "innovate in architectural forms with the use of local cultural and post-modern symbols".

Built on a small site, the building is shaped like a giant cubic with a protruding part on the 21st floor, a striking microwave tower over the top and sloping roof, all of which were bold, innovative treatments back in the late 1980s.

The architects made efforts to allow the high-rise to mingle with the historical context of Nanjing Road. Triangle windows, inspired from the old Shanghai dormer windows, and pentagon windows that relate to Gothic arched windows, were adapted to echo with the Bund nearby and the adjacent Trinity Church on Jiujiang Road.

Since its competition in 1988, the building has been widely discussed and studied by scholars, and won many architectural awards, including one of "Shanghai's 10 Best Buildings" in 1989, "Shanghai Classic Building" in 1999 and a big award for creative design by the Architectural Society of China in 2009.

But people continued to have different opinions about the building. A 1990s survey showed most architects and architectural scholars had positive comments about the building, while up to 53 percent of the general public thought it was an ordinary design. At least six percent of them thought it was an ugly building.

Shanghai-born artist Zhang Lansheng is one of them. "During that era it was a very modern building, but I couldn't admire it aesthetically. The post-modern feel has faded over the years. It's no longer eye-catching in a city which is filled with post-modern, futuristic buildings. However, it mirrors a special time which is probably the value of preserving it," says Zhang, who was a professor of art history in Shanghai and later in Sydney, Australia.

In 2013, Shandong-based real estate and new energy developer Luneng Group took over the building and the adjacent No. 181-a 1930s Art Deco building which was the headquarter of Shanghai Power Company.

The developer invited four architectural firms, Chinese and foreign, to come up with a plan to merge the two buildings into a luxurious boutique Edition Hotel. The plan was part of the local government's efforts for urban regeneration, lack of upper-class hotels along Nanjing Road E. and Luneng's purpose of branding itself on a prominent location in Shanghai.

Most plans changed the facade or wrapped it with another facade. Another plan was to convert the original modern facade into an Art Deco one with vertical lines to go with the 1930s Art Deco building.

It sparked controversy, heated discussion and media attention after Tongji University professor Lu Di posted about it on his Weibo site. The Architectural Society of Shanghai even organized a "quartet dialogue on urban regeneration" in 2016 to discuss the fate of this building. It was attended by representatives from the government, the developer, the designing firms and the media.

After many discussions, the plan was finally altered and emphasis was placed on protecting the historical memory of the building that allowed for changing the interior but required 90 percent of the facade to be preserved — a very high rate. All the four major elements: the silhouette, the triangle windows, the sloping roof with the microwave tower and the architectural color had to be preserved. An atrium was added to link the building with the No. 181.

During the "quartet dialogue", Chen Haitao of the Luneng Group said the principle of the project was "form follows function" but the process was "very difficult" .Four architectural firms designed more than 50 different plans to "save the building" .

"It's a challenging project because the building's regeneration has a lot of limitations. We have to maintain the original size and the original elements. Meanwhile, we needed to change the function from an office building to a hotel. It's like stuffing a peach into an apple," says Fan Jiashan of Chinese company ECADI, chief architect of the project.

"People are happy to see the facade has been successfully preserved but they don't know the backbreaking process. The developer has also spent a lot of time, energy and money on it," Fan adds.

In a public lecture, Tongji University professor Hua Xiahong compared the project with the 888 Julu Road case, which attracted wide attention when the owner used steel and concrete structure to replace a 1920s villa designed by Slovakian-Hungarian architect L. E. Hudec without permission as "it was too dilapidated to be repaired". The owner was fined 30.5 million yuan (US$4.46 million) for damaging a historic villa.

Hua says that the two buildings, 201 Nanjing Road and 888 Julu Road, were built in different styles in different eras, but "they encountered similar problems during renovation and mirrored the attitudes of different powers towards conservation in our city".

"For conservation, the city needs to have its own character. In the 1990s people preferred newer things in China but nowadays more people are realizing the value of history. The two buildings are interesting cases of the process," she says.

Yesterday: East China Electrical Power Building   **Today:** The Shanghai Edition Hotel

Address: 201 Nanjing Road E.   **Date of construction:** 1988

Architects: Luo Xinyang and Qin Yong   **Architectural style:** Modern

Tips: Please admire the facade and the four major elements that have been preserved after considerable efforts — the silhouette, the triangle windows, the sloping roof with the microwave tower and the architectural color.

"旗舰店" 大楼

A Building for Flagships

南京东路257号。不同时代里，知名品牌相继把"旗舰店"开设在这座位于南京路黄金转角的大楼里。

第一个品牌是老上海丝绸大王"老介福"。1860年，老介福由一对来自福建的祝氏兄弟创立于九江路，原名介福。祝氏兄弟后来退出生意，把店铺转让给苏州人程芦舟。

1936年老介福迁到现址，成为上海最大的丝绸商店，不仅销售高档丝绸，还根据从海外收集的时髦图案开发设计自己的产品，赢得"中国丝绸之王"的美名。除了经营丝绸，老介福也提供呢绒产品和定做服装的服务。

同济大学钱宗灏教授推测，这座大楼的建造可能与南京路、河南路的拓宽有关。1935年英文《大陆报》提到在上海工部局的一个市政项目中，该地块的老建筑被拆除，南京路和河南路得以顺利拓宽，印证了钱教授的猜想。报道还提到，哈同洋行在此投资兴建高达6层的哈同大楼，由建筑师德利（Percy Tilley）设计。这位建筑师主持设计了哈

同洋行不远处的另一个项目——嘉陵大楼。哈同大楼沿南京路河南路的转角处理为弧线。

"大楼底层立面将用黑色大理石贴面，上部是砖砌墙面。楼里设有商店、办公楼和小型公寓。"报道写道。

老介福将旗舰店设在这座新大楼里。老上海英文报纸的购物栏目经常推荐这个"丝绸宫殿"的产品和服务。1938年，《大陆报》派一名记者体验采访"丝绸宫殿"。一走进南京路257号，记者的眼睛就被40英尺宽的绉纱吸引了。绉纱的设计明快，图案是多彩的碎花。

"不过，那种印象迅速消散了，我看到无穷无尽的丝绸、缎面、花色绉纱，感到眼花缭乱。今年最时髦的东西，都唾手可得。"记者写道。

这位记者发现面对这么多诱人的商品，难免有选择困难症，不过店里的服务一流，"客人们可以悠闲地选购商品——就算一直挑选，或不停地改主意，哪怕购物长达数小时，也没有人会介意。"

"除了花色丝绸，店里还有用来制作端庄晚装长裙的亚光白色丝绸。此外，还有彩色或黑白条纹的丝绸，后者最适合那些有智慧的女人，她们非常确信一套黑白套装赋予自己的成熟魅力。而微皱的绉纱无需熨烫，适宜旅途，能让人在一整天下来仍看起来衣着整齐。总而言之，无论什么样的丝绸，老介福都有售，"记者写道。

除了零售，老介福还为华懋饭店（今和平饭店）等高档酒店提供窗帘、沙发套和床单等软装用品。这些在苏州定制的专供品质量很高，英国著名演员卓别林20世纪30年代访沪时专门到老介福订购了几打丝绸衬衫。南京路上的这家丝绸大王因为货品花色多、质量高、营销手段多，生意越做越大，在呢绒丝绸业独占鳌头。

同济大学常青院士认为老介福大楼和沙逊大厦、惠罗公司一样，是南京路的城市地标，已经成为市民城市记忆的一部分。南京路上的两家百年老店——老介福和亨达利钟表是一中一西两个传统品牌的代表，因为其卓越的品牌影响，已经融入上海市民的日常生活中。

在老介福生意最红火的黄金时代，一家历史悠久的外资企业——西门子中国分公司也入驻了这座大楼。1937年《大陆报》的报道提到，位于哈同大楼四楼的西门子办公室装修得现代摩登，在商业区显得非常养眼。公司在大楼底层还设有展示厅，展示西门子种类繁多的商品。

"装修的基调是海绿色、象牙色与黑色。柜台和桌子都是简洁现代的款式。吸引来宾注目的是一面镶嵌白框的墙面，展示了关于照明设备和电炉等产品的广告，设计富有艺术的气息。房间里还展示着其他一些设备，如石英台灯、尺子、布线设备，展陈布置品位不俗。"报道写道。

1949年后，老介福转变为国营单位，这家公司在南京路的命运也随着时代变迁慢慢改变了。老建筑在1993年和1999年两次大修，2017年黄浦区政府对南京路东段进行城市更新之际，外立面又历经修缮。

2012年,《纺织装饰科技》刊登了一篇题为《"老介福"翻牌"快时尚"》的文章。报道提到,为了实现更加年轻化的发展,老介福商厦选择与美国快时尚品牌"Forever 21"携手,开设上海首家旗舰店,也是该品牌全球最大的旗舰店之一。当年这家旗舰店的开设被认为是美国中小企业进入中国市场的象征。老介福在新商场曾辟出工作室,为老顾客量体裁衣,后来慢慢淡出了南京路。

时过境迁,快时尚行业后来陷入瓶颈,2019年Forever 21退出中国市场后,这座引人注目的大楼又迎来引人瞩目的新租户。2020年6月24日,由华建集团历保院修缮设计的华为旗舰店在此开幕。这是华为全球最大的旗舰店,占据大楼一到三层,面积达5000平方米,成为南京路新的"城市客厅"。

铁打的南京路,流水的旗舰店。从传统中华丝绸到美国快时尚再到民族科技品牌,哈同大楼百年商业的戏剧化更迭,好像一面镜子,映射了中外品牌时代变迁中的机遇和挑战。

昨日:哈同大楼　**今天**:南京大楼（华为旗舰店）　**地址**:南京路233-257号
建造时间:1935年　**建筑师**:德利洋行 Percey Tilley
参观指南:建筑内部历史旧貌不多,了解建筑历史后再漫步这座"旗舰店大楼",别有一番滋味在心头。

The historical building at 257 Nanjing Road E. has served as a flagship store for famous brands in different eras. It was once home to an upper-class silk emporium, the Laou Kai Fook & Co. Ltd., in the 1930s and became the first Shanghai store of American cheap-chic retailer Forever 21 in the 21th century. China's leading technology company Huawei decided to open its largest flagship on earth in this building that perches around a golden corner of Nanjing Road.

The first brand Laou Kai Fook was originally named Kai Fook and was established in 1860 on Jiujiang Road by the Zhu brothers, who were from Fujian Province. They later sold the shop to Cheng Luzhou of Suzhou.

The store later moved to No. 257 Nanjing Road E. and grew to become the city's largest silk shop, which not only sold top-quality silk products, but also designed popular patterns. The shop came to be known as "The King of Chinese Silk." It also sold wool fabrics and offered tailor-made services.

Tongji University professor Qian Zonghao says the building came up following the widening of Nanjing and Henan roads.

A news story in *the China Press* in 1935 proved his assumption. The report says the destruction of the old buildings on the property would allow the widening of Nanjing and Henan roads in accordance with the policy of the Shanghai Municipal Council. The facade of the building had a broad curve at the corner, just like many of the newer buildings back then.

The report also gave a description of the six-story building that was designed as S. A. Hardoon Building for the Hardoon Estate by architect Percey Tilley.

"The building will have black marble on the first floor, while the upper stories will be of brick. It will contain stores, offices and small apartments," it said.

Laou Kai Fook turned the shop into a paradise of silk goods. In 1938, *the China Press* sent a reporter to write about the shop whose eyes were struck upon entry into the silk store at 257 Nanjing Road E. by "a 40-inch-wide crepe remaine, completed served with the brightest design of multi-colored, tiny flowers."

"But that impression is soon lost, and one is left bewildered by the endless variety of silk and satin, flowered crepes, the height of fashion this year, are available by the tableful," the reporter described.

The reporter noted the difficulty of choosing among so many attractive materials. However, the shop's service was remarkable, whose staff "requests clients to be at their leisure — no one minds if a customer remains choosing and constantly changing orders for hours on end."

"Apart from the flowered things, there are delicate matt white silks, which would make up into the

most dignified gowns for formal evening wear — or striped silks in colors or cool blacks and whites, these latter for wise women who recognize the immense certitude and sense which a coolly reserved black and white ensemble gives to maturity. Crinkled crepes, kindly creatures who wouldn't know the use of an iron if they saw one, suggest prettily that they would like to accompany you on your travels, or point out that you can look as fresh at the end of the day as at the start. In short, say 'Everything in Silks' and you say Laou Kai Fook," the reporter wrote.

Laou Kai Fook & Co. also made textile goods for the Cathay Hotel, including curtains, slipcovers, bed sheets and other textile decorations. All these exquisite textile goods were made from high-quality silk of Suzhou within half a year. Even famous British actor Charlie Chaplin purchased dozens of silk shirts from the shop when he visited Shanghai in the 1930s.

"The building has become one of the landmarks as part of Shanghai's urban memory, along with the Sassoon House and Whiteway Laidlaw Building on Nanjing Road," says professor Chang, author of the book "Nanjing Road -- Origin of a Metropolis".

"Laou Kai Fook and Hope Brothers which sold clocks and watches were two traditional brands on Nanjing Road. One Chinese and one Western, they have merged into daily life of Shanghai people for a very long time," he says.

While Laou Kai Fook was having good business at 257 Nanjing Road E., another century-old company, Siemens China, opened a new office and showroom on the fourth floor of the same Nanjing Road edifice.

According to *the China Press* in 1937, the offices had been completely decorated and furnished in modern style, offering a pleasing note in business quarters. The showroom on the ground floor adjoining the building entrance displayed the company's varied wares in a striking setting.

"The color scheme of the decoration is sea green, ivory and black, with counters and desks of simple, modern style. The visitors' attention is drawn to a white colored frame extending the entire length of one wall, containing artistically executed advertisements for lighting fixtures, irons, electric stoves and other appliances, while quartz lamps, meters, wiring de-

vices and countless other items are tastefully displayed throughout the room," the report described.

In the 1930s, Siemens China was a branch of the German Siemens Works and maintained offices in today's Guangzhou, Wuhan, Hong Kong, Nanjing, Beijing and Tianjin.

After 1949, Laou Kai Fook became a state-owned company but its fate on Nanjing Road has changed during the times. The building was renovated twice in 1993 and 1999 and later faced a big renovation when Huangpu District was upgrading business in the eastern section of Nanjing Road.

According to a report on "Textile Decoration & Technology" in 2012, to house young customers of the east section of Nanjing Road, Laou Kai Fook retreated to smaller shops in communities and maintained only a tailor-made studio in the former department store which had been renovated to be a big flagship of American fashion retailer Forever 21 Inc.. The four-story branch of the Los Angeles clothing maker in the 1930s silk emporium was one of its largest flagships around the world and was regarded as a symbol that middle and small American enterprises entering the attractive Chinese market.

Now both Laou Kai Fook and Forever 21 have vanished from this prominent building on Nanjing Road. China's leading technology giant Huawei opened a grand 5000-square-meter store, its largest flagship on earth here. For some time, logos of Laou Kai Fook, Forever 21 and Huawei co-exist on the façade of this flagship building.

The dramatic change of flagships in this Nanjing Road building is a mirror of Chinese and foreign brands, old or new, fronting the change and challenges of times.

Yesterday: S. A. Hardoon Building   **Today:** Nanjing Building (Huawei flagship store)
Address: 233-257 Nanjing Rd. E.   **Date of construction:** 1935
Architect: Percey Tilley   **Tips:** Though the building's interior has been changed greatly, it's interesting to visit it after knowing its history behind.

"老介福"延续丝绸传统
Laou Kai Fook perpetuates silk tradition

您还记得"丝绸裙和衬裙"的时代吗？女人为丝绸摩擦发出的沙沙作响声而感到自豪，这种声音让她们感到一种只有丝绸才能带来的穿着考究的优雅感。但是衬裙的使用范围更大，尽管我们现在又开始在沉湎于塔夫绸面料，"丝绸沙沙作响的声音已经减弱了"。如今，我们正在寻找可以轻柔地垂入连衣裙的面料，这样我们就能穿着裙子安静地行走。

我们必须跟随潮流，但丝绸是一种多样性的面料，可用于各种式样和类型的衣服。它是强度最高的纺织面料，其弹性有助于服装保持挺括形状。在九江路250号的"老介福"商店，你可以找到所有类型和等级的丝绸产品。购买前先确定它们的外观，然后再决定是否要用这块丝绸来制作圣诞节的连衣裙。

几个世纪以来，颜色和图案优美的中国丝绸为生活带来魅力和浪漫气息。老介福秉承这一传统，提供各种丝绸供您选择。

他们会告诉您，一个古老的传说将丝绸文化的起源归因于一位中国皇后。她培育了蚕种，将蚕茧中的蚕丝纤维卷起染色并织成有光泽的织物。然后她用这种织物为皇帝制作了一件特别美丽的礼服。那是4000多年前的事了。故事就这样讲的，子孙后代们把这位高贵的女士奉为"丝绸女神"。

今天，每个衣橱里都有丝绸的礼服。丝绸是必须购买之物，因此在老介福商店购买丝绸时，你会发现各种档次和价格的丝绸产品，从塔夫绸到有光泽的缎子再到天鹅绒和锦缎，应有尽有。

Do you remember the days of "silk skirts and petticoats"? Women were proud of that rustle which gave them a feeling of well-dressed elegance that only silk could give. But petticoats have turned out to a greater extent and though we are now learning to luxuriate in taffeta again, that "silken rustle has been softened." Nowadays we look for textures that will drape softly into frocks in which we can walk silently.

We must follow the mode, but silk is a versatile fabric lending it to every mode and every type of dress. It is the strongest of textile fabrics, and its elasticity helps garments to keep them shape and to resist erasing. The better types and grades of all silks are to be found at Laou Kai Fook at 250 Kuikiang Road, and by all means, give their things a look before definitely deciding on that Christmas frock.

For centuries, Chinese silks of lovely colors and designs have lent glamour and romance to life. And Laou Kai Fook is carrying out that tradition, offering you an amazing array of silks from which to choose.

An ancient legend, they will tell you, attributes the beginning of silk culture to a Chinese empress who cultivated the silkworms, reeled the silk fibers from their cocoons, then dyed and wove them into a lustrous fabric, with which she made it a superbly beautiful ceremonial garment for the emperor. That was over 4,000 years ago, so the story goes, and posterity has deified this noble lady as the Goddess of Silk.

Today, there are silk gowns in every wardrobe.

If necessity you must shop for silk, so when you do so at Laou Kai Fook's you will find a wide range of qualities and prices in all sorts of silks from taffeta and lustrous lengths of satin to cut velvet and brocade.

摘自 1934 年 12 月 13 日《大陆报》
Excerpt from *the China Press,* on December 13, 1934

骄傲的国货公司

An Ambitious Emporium

南京路曾经是中国版的纽约"第五大道",可并非这条路上的所有商店都生意兴隆。1931年,宏伟的大陆商场在初建时貌似是一项成功投资,后来却未能给业主大陆银行带来丰厚回报。

大陆商场是一座回字形布局的西方办公楼,中央有大天井。这座由中国建筑师庄俊设计的大楼呈现简洁的装饰艺术风格。大楼初建时,英文《北华捷报》称这是一座"百万美金大楼",是一个"南京路上充满雄心的计划"。

"大陆银行新成立的信托部近期决定将一个新颖而耗资巨大的项目付诸实施。这个项目不仅会让南京路的美丽增色不少,也会满足当地消费者的需求。"1931年4月14日的《北华捷报》报道写道。

报道介绍,这座高达7层的巨厦内将会开设一家大型购物中心,与北京一家著名的东城市场类似,但规模更要大。大陆商场占据了南京路的一个街区,商场内还设有写字间、饭店和一家现代旅社。建筑面朝南京路的正立面超过300英尺宽(约92米)。建筑师设计了一个T字型的内部通道,便于车辆驶入。

大陆商场项目酝酿于上海的黄金年代,这段从20世纪20年代到30年代初的岁月是上海近代经济发展最繁荣的时期。南京路的商店挤满了顾客,进口商品常常售罄。而在上海营业的银行很多都资金雄厚,急于将多余资金投资到股票、债券和房地产项目中。上海很多高层建筑都是这段时期由金融机构投资兴建的,如四行储蓄会投资的国际饭店和万国储蓄会建造的毕卡迪公寓(今衡山饭店)。

投资大陆商场的大陆银行是1923年成立四行储蓄会的"北四行"之一。20世纪20年代末,大陆银行的存款和信托总金额超过1000万元,银行创始人兼总经理谈荔孙对投资上海红火的房地产市场开始动心。他计划在南京路黄金地段买地建楼,开一家专卖国货的商场,与南京路四大百货竞争。为了宣传大陆银行的品牌,他将新厦命名为"大陆商场"。谈荔孙的团队筹备项目时

做了认真测算，根据当时上海蓬勃发展的经济和房地产市场，认为未来投资的收益可观。

大陆银行从拥有大量南京路地产的哈同洋行租下基地，委托庄俊设计新厦。庄俊出生于1888年，于1910年和1923年先后赴美国伊利诺伊大学和哥伦比亚大学学习，1925年回上海开设了自己的设计事务所。庄俊早期作品多为西方古典主义风格，如金城银行（今江西中路200号交通银行）。

同济大学副校长伍江认为，庄俊的设计水平媲美同时期的西方建筑师，但他在设计大陆商场时开始风格的转变，彻底摒弃了古典主义装饰，采用装饰艺术风格的图案，开始倾向设计简洁的建筑外观。而庄氏此后的作品，

特别是延安路上的孙克坚产科医院呈现出更现代的"国际式"风格。

庄俊还发起成立中国建筑师协会，以此来团结中国建筑师与外国建筑师竞争。这位资深建筑师也以长寿而闻名，他喜欢多种运动，92岁还洗冷水澡，96岁时坚持每天散步，102岁高龄时去世。

大陆商场的项目虽然经过精确测算和精心设计，投资也很巨大，却没有成为一桩成功的生意。1932年"淞沪抗战"和1935年"银元危机"接连爆发后，南京路繁荣的景象不再，商铺租赁价格下跌，对于新建成的大陆商场来说租赁更加困难。

上海金融史专家邢建榕认为这座雄心商厦定位尴尬。"南京路四大商场提供时尚高档的进口环球百货，如法国化妆品、瑞士钟表和德国电器用品。大陆商场对大楼分割出租，招商经营，一间间中小商铺分布其中。其货色定位中档，一般市民买不起，富裕阶层不愿来。大楼里的办公空间因为与商场同处一楼内，也不够体面，入驻的多是中小企业，少的时候房间租出只有五六成。"他说。

1949年后，大陆商场改为国

营商店，曾开过一家新华书店的旗舰店。如今，新华书店仍在这座大楼里，只是店面大幅缩小。

大楼经过改建，把昔日小店铺的格局打通，改为一家购物中心。英国娱乐公司Merlin Entertainment在大楼里开了一家名为上海惊魂秘境"Shanghai Dungeon"的沉浸式剧情体验项目。惊魂秘境使用高科技手段和真人演员，将游客带回老上海，通过大约10个"黑历史"和恐怖传奇，展示上海"色彩斑斓的过去"。

邢建榕还研究发现，大陆银行为大陆商场项目共投资了超过370万元，但到1937年仅收回了100万元。银行不得不忍痛把商场低价卖给哈同洋行，以结束这个失败的项目。

"不知道是否与投资失败有关，谈荔孙变得性情抑郁，不久因病逝世。在社会转型、政局动荡不定之际，银行的房地产投资，也暗伏着杀机，充满着风险。"他总结道。

历史是一堂最好的课。大陆商场本身的历史，也许会为在这座大楼里每日上演的惊魂秘境，增添一些惊心动魄的味道。

昨日：大陆商场　**今天**：悦荟广场　**地址**：南京东路 353 号
建筑师：庄俊　**建筑风格**：装饰艺术风格
参观指南：请细细欣赏外立面上由一位中国建筑师设计的装饰艺术细节。

Nanjing Road had a reputation as "China's Fifth Avenue," but not all commercial establishment on the street enjoyed success. The grand Continental Emporium, built in 1931, seemed to be a promising project at first but failed to reap in the profits for its investor, the Continental Bank.

"The emporium was designed in the shape of a modern Western office building. It was enclosed by walls with a large yard inside with cars driving into the building. The facade displayed a simple Art Deco style but essentially had only two Art Deco elements, the tower and the vertical lines," says Tongji University professor Qian Zonghao, author of the book *Nanjing Road (1840s-1950s)*.

Back in 1931, the emporium was called a "million-dollar building" and "ambitious scheme underway on Nanking Road" by *North-China Herald*.

"The newly established Trust Department of the Continental Bank has recently decided to materialize what may be considered a very novel and ambitious project that promise not only to enhance the beauty of Nanking Road, but also to meet a real demand of the local populace," the newspaper reported on April 14, 1931.

The report said the 7-story building housed a large emporium, which was similar but more gigantic than the famous East City Mart in Beijing. The project took up a whole block on Nanjing Road with spaces for office rooms, restaurants and a modern hotel.

The frontage on Nanjing Road was more than 300 feet (90 meters). In order to enable automobiles to reach any part of the building, a T-shaped lane was constructed within the premises.

The project was initiated during the "golden era" of Shanghai — from the 1920s to early 1930s — when the economy was booming. Big department stores on Nanjing Road were flooded with customers and imported products would sell out quickly.

With soaring deposits, local banks were eager to invest the funds in bonds, stocks and real estate. Many of the city's well-known high-rises were built during this period by financial institutions such as the Park Hotel by Joint Savings Society or the Picardie Apartments (today's Hengshan Picardie Hotel) by International Savings Society.

Continental Bank was one of the four big northern Chinese

banks that formed the Joint Savings Society, a cooperative venture and an influential Chinese financial institution founded in 1923. With deposits and trust savings of more than 10 million Chinese dollars in the late 1920s, Tan Lisun, founder and general manager of Continental Bank, couldn't resist the temptation of investing in Shanghai's booming real estate market.

He had this ambitious plan to build a department store at a great location and nice square shape on Nanjing Road to sell Chinese products and compete with the four big Chinese department stores on the street. To boost the reputation of Continental Bank, he named the building Continental Emporium.

The project appeared to be a smart investment by Tan and his team as both the city's economy and real estate market were soaring. The bank commissioned Chinese architect Zhuang Jun to design the building after renting the site from Jewish tycoon Silas Aharon Hardoon who owned many properties along Nanjing Road.

Zhuang went to study architecture in Illinois University in the United States in 1910, and later at the Columbia University in 1923. He established his own design studio in Shanghai in 1925.

Among a galaxy of Chinese architects in old Shanghai, Zhuang was the first to study abroad and also the first to return home and practice architectural design in China.

Wu Jiang, vice president of Tongji University, says Zhuang initially preferred Western classical style, like his noteworthy work, the Kincheng Bank (today's Bank of Communication at 200 Jiangxi Road M.). But Continental Emporium marked a change of style in which he thoroughly abandoned classical decoration, adapted Art Deco patterns and designed a simple-cut architectural appearance.

Zhuang's later work such as the Sun Ke Jian Maternity Hospital showcased a more modern "international style". He also founded the Society of Chinese Architects to unite the Chinese architects in a market with foreign architects. This architect lived until the age of 102.

But the Continental Emporium he designed took a hit after the Japanese army invaded Shanghai in 1932 and the "silver crisis" broke out in 1935. Nanjing Road was no longer as prosperous as

before. The Continental Emporium struggled to lease out much of its shopping space on the three floors. The top floors were used as offices.

"Compared with the four big department stores on Nanjing Road which offered the most fashionable, upper-class imported products like French cosmetics, Swiss watches or German instruments, the Continental Emporium was filled with small-sized shops that sold middle-level products, which failed to attract the wealthy families and were also out of reach of the ordinary citizens. The offices, too, did not have a decent environment," Shanghai historian Xin Jianrong says.

After 1949, the gigantic building housed state-owned shops and later opened the flagship Xinhua Book Store.

Today, the big bookstore has retreated to a much smaller shop in the building, which is again a shopping mall featuring shops, restaurants, entertainment and an entertainment project named "The Shanghai Dungeon."

The project operated by Merlin Entertainment from UK. uses themed sets, live actors and cutting-edge technology to take visitors on a journey back in time.

Xin says the Continental Bank spent more than 3.7 million Chinese dollars on the project back in 1937 but only recovered 1 million. The bank had to sell the department store to Hardoon's company at a low price.

"I'm not sure if this failure had anything to do with it, but manager Tan became depressed and died of illness soon after. At an era of social transformations and unstable political condition, investment on real estate was always full of risks and dangers," Xin says.

History is often a good lesson. The history of the emporium itself

has added a special color when the Merlin recreates the city's colorful past.

Yesterday: The Continental Emporium   **Today:** The Mosaic Mall

Address: 353 Nanjing Road E.   **Architect:** Zhuang Jun

Architectural style: Art Deco   **Tips:** Please admire the simply cut Art Deco facade designed by the renowned Chinese architect.

大陆银行大楼
A Continental landmark on Jiujiang Road

大陆银行在计划创办南京路大陆商场时，在邻近的九江路建起了新的银行大楼。大陆英文《大陆报》称为"另一座价值数百万美元的中国建筑"。

高达10层的大楼由基泰工程司设计，是一幢气势宏伟的现代建筑。根据1933年12月6日《北华捷报》报道，银行大厅巧妙地使用了大理石装饰，简洁而庄重。

"外立面的底部是来自青岛的熟练工匠用青岛花岗岩铺设的。经理套房和会议室呈现了中国宫殿风格，操刀的画家和工匠曾经为前朝的皇室工作过。一间房间的天花板是由古老的中国钱币组成的设计，展示了钱币的发展。而门、门框和照明装置的铜制五金件是由中国工匠根据建筑师设计悉心锻造的。"

如今，大陆银行大楼由另一家中国金融机构——上海信托作为总部使用。近年的修缮改造工程恢复了中式风格的房间和由美国莫思勒公司（Mosler & Safe）制造的四个保险金库。

"保险柜是大陆银行的一项重要业务，该大楼保存了当年留下的1,180多个精致的保险柜。20世纪90年代我们搬入大楼后，公司还开展过保险柜租赁业务。现在有50多个保险柜的用户联系不上，而这些保险柜中的贵重物品仍然保存完好。"上海信托市场营销部总经理浦剑悦说。

根据《大陆报》报道，位于银行大楼夹层的大型保险库是最令人印象深刻的特征之一。银行管理层、设计建筑的基泰工程司、金库制造商美国莫思勒公司和其上海代理商慎昌洋行（Anderson Meyer Co. Ltd.）密切合作，"结合了现代金库建筑和建造所有最新和最好的特色"，让这个保险库典范成为现实。

保险库的主体由18英寸厚的钢筋混凝土墙和选用最高等级的钢铸造成的钢衬组成。当入口和紧急门处于关闭位置时，它成为一个气密

室。保险库门由实心铸钢和一系列特殊钻孔的防烧板制成。它由24个直径3英寸的钢螺栓锁住。这是国内最坚固的保险库门之一，可以防范袭击。

While planning the grand emporium on Nanjing Road, the Continental Bank built a new bank and office on neighboring Jiujiang Road which was also called "another million-dollar Chinese building" by the China Weekly Review.

Designed by Messrs Kwan, Chu and Yang, the 10-story building was an imposing modern structure. According to a December 6, 1933 report in the North-China Herald, the main banking hall boasted simplicity with marble judiciously used.

"The lower part of the exterior is executed in Tsingtao granite by skilled workmen brought down from that region. The manager's suite and the conference room are finished in Chinese palace style by painters and craftsmen who worked for the Imperial House in the late dynasty. The ceiling panels in one room are designs composed of old Chinese coins illustrating the development of coinage. The bronze work of doors, frames and lighting fixtures have been carefully wrought by Chinese workmen from the architect's designs," the report described.

The building now serves as the headquarters of another Chinese financial institution — the state-owned Shanghai Trust. A renovation has restored the Chinese-style chamber with exquisite ceiling panels and four safe vaults built by American company Mosler Safe.

"Safe was an important business of the Continental Bank, and the building has preserved more than 1,180 delicate safes left from the era. Our company also conducted safe rental business after we moved into the building in the 1990s but lost contacts with some customers. We still couldn't find the owners of 50 safe boxes whose valuables are still well-preserved," says Pu Jianyue, general manager of Shanghai Trust's market department.

According to the China Press, the massive deposit system on the mezzanine floor of the bank building was one of the most impressive fea-

tures. Close cooperation among the bank management, the Architects Messrs Kwan Chu & Co., the manufacturer Mosler Safe Company and the local agent Anderson Meyer Co. Ltd. have made a fine example of safe deposit construction, "incorporating all the latest and finest features of modern vault architecture and construction".

The body of the vault is constructed of specially reinforced concrete walls that are 18 inches thick, supplemented by a steel lining composed of the highest-grade steel carefully fitted together. It becomes an airtight chamber when the entrance and emergency doors are in closed position.

The door is constructed of solid cast steel and a series of special drill and burn-proof plates. It is locked by 24 steel bolts that are 3 inches in diameter — one of the strongest vault doors in the country and is attack-proof.

"四大公司时代"的序幕
Sincere Started the Era of Four Big Stores

1917年10日20日，南京路先施公司开门营业，顾客盈门。先施公司是"南京路四大公司"开业的第一家。这一天开启了南京路商业的新时代。此后20年间，其余三家百货公司——永安百货、新新公司和大新公司——相继开业，"四大公司"既激烈竞争，又交相辉映。

先施公司开业的前一天，英文《上海泰晤士报》介绍了公司的发展历史。1900年，先施公司创立于香港，后在广州和新加坡开设分公司。报道写道，"人们渴望的所有商品——眼镜已为消费者备好，手表修理业务已开张，来自多国的商品陈列得相当诱人，由花环和棕榈树装饰衬托着。咖啡馆、剧场和电影都为迎接顾客光临而准备就绪。"

上海档案馆研究员张姚俊翻开馆藏的"四大公司"档案，发现它们的发家史有相似之处：先施公司老板马应彪、永安公司创办者郭氏兄弟、新新公司创始人之一李敏周和大新公司控股人蔡昌都来自地处珠江出海口的香山县（今广东中山）。他们早年在澳洲经营果栏（即水果批发店）或杂货店起家，然后带着实业救国的理想回国二次创业。

"尽管先施、永安和大新公司均在香港和广州取得斐然业绩，但马应彪等人不约而同地把目光瞄准'殷商巨贾咸集'的上海，选择在当时申城地价最高的南京路落地生根，这不能不说是'英雄所见略同'。"张姚俊说。

1914年，先施公司创始人马应彪曾亲自到南京路调研。他发现南京路靠外滩一带都是外资百货商店，而西侧则开着很多狭小的中式店铺。这些店铺的商品种类有限，无法与外资百货公司竞争。调研让他下定决心，在这条著名商业街上开一家大型百货公司。

"马应彪选址谨慎，他仔细分别计算路南和路北的人流后，挑选位于南京路浙江路口的位置。先施公司的原址是一家茶楼点心店。"同济大学钱宗灏教授说。

马应彪从英商业主租下这块地皮后，盖起一座7层高的新古

123

典主义风格大楼。

先施公司的英文名"Sincere"取意为"货真价实"。

1917年10月20日英文《北华捷报》报道，先施大楼是钢筋混凝土结构，以砖石筑成。商店中央有带装饰的天窗，宛若一个"采光井"。商店有两座楼梯，外墙饰有蓝色宁波石、日本花岗岩和水泥，柱式和檐口等用人造石装饰。

"这家商店是上海华资百货商店中规模最大也是最先进的，货品从缝衣针到大象玩具都有售。底层是食品和日用杂货，包括葡萄酒、烈酒、厨房用品、五金器具、雪茄、烟草、吸烟用品、药品、香水、糖果等。二层则销售布料、女帽、订制服装、男装、皮草等。三层是瓷器、电器、运动用品、玩具、乐器等，四楼是公司办公室和家具部。"报道介绍。

钱宗灏教授认为，先施大楼这座建筑以西方新古典主义风格为主，局部有巴洛克装饰，是一座带塔楼和柱饰的商业建筑。

作为南京路上第一家大型的华资百货公司，先施公司经营颇为成功。马应彪和他的团队使用新奇手段促销，如设立屋顶花园和刊登广告招聘女营业员。后者是大胆的创新，因为百年前传统中国女性多在家相夫教子，很少出门抛头露面。招聘广告刊出后无人应征，马应彪的夫人霍庆棠主动请缨担任化妆品售货员，开创了先河。后来，女营业员们成为先施公司最好的广告。

沈寂主编的《老上海南京路》一书提到，先施、永安、新新、大新四家中国人经营的百货公司规模远大于英资百货公司，顾客以国人为主，生意日渐红

火，而英资百货公司日渐衰落，四大公司成为南京路的标志。

如今，先施大楼由锦江之星酒店和上海时装商店共用，后者是上海百联集团旗下的公司，以销售羊绒衫和中老年服装为主。其他三家也都是国营商店，新新百货是第一食品商店旗舰店，大新公司（第一百货）和永安百货在南京路新一轮城市更新中修缮改造，力图吸引新时代的消费者，再现昔日荣光。

"'四大公司'不仅联袂开创了上海近现代百货业发展的新纪元，更是当年引领这座城市摩登生活的风向标，改变了上海人的消费观念与生活方式。如今在南京路上，'四大公司'的巍巍高楼依旧矗立，而记录'四大公司'历史的档案仍在上海市档案馆完整地珍藏着。" 张姚俊说。

昨天：先施公司 **今天**：上海时装商店 & 锦江之星酒店 **地址**：南京东路 690 号
设计师：德和洋行 **建筑风格**：西方新古典主义
参观指南：建筑作为上海时装商店对外开放。

The doors of Shanghai Sincere Department Store were thrown open to the public 100 years ago that launched the era of "four big Chinese department stores on Nanjing Road."

"The usually quieter corner of Nanjing and Zhejiang roads was bustling on October 20, 1917 with curious locals visiting the newly opened Sincere Department Store. Their eyes dazzled at the sight of global products and the store was packed with people until evening," says Shanghai Archives Bureau's Zhang Yaojun, who did extensive research on the four big department stores on Nanjing Road.

"Within two decades, three other big department stores — Wing On, Sun Sun and The Sun — sprung up successively on the golden Nanjing Road," he adds.

A day before the opening of Sincere store, *the Shanghai Times* published a story tracing the history of the Sincere Co. back to 1900 in Hong Kong and later in Guangdong and Singapore before reaching Shanghai.

The report also previewed "everything that the heart of man or woman desires: the optician is ready to test the eyes of a customer; the watch repairing estab-lishment is open; while the wares and products of many countries are displayed attractively amid a wealth of flowers and palm trees. The cafe, the theaters, the cine-matographs, etc, are bound to attract the crowds ..."

Zhang has found considerable similarities among the four department stores whose founders were mainly Chinese merchants, all of whom earned their first buckets of gold by selling fruits in Australia. They returned to China with the intention of "saving China through business".

"Sincere, Wing On and The Sun companies had already made remarkable achievements in Hong Kong and Guangzhou, but these Chinese entrepreneurs had cast their eyes on Shanghai, especially Nanjing Road, leading to a surge in land prices back then," Zhang says.

Ma Yingbiao, founder of the Sincere Co. selected the venue after conducting a careful field survey of Nanjing Road in 1914.

Ma found that the section of Nanjing Road E. near the Bund was dominated by big foreign-owned stores like Hall & Holtz and Whiteway Laidlaws. However, the section west to Henan Road was crammed with

small Chinese shops, which had limited products and could not win against their foreign competitors.

The survey strengthened Ma's determination to open a big Chinese-owned department store on the prominent Nanjing Road.

"After the survey of the location and people, Ma handpicked the site to build the Sincere store. He chose the former site of a dim sum shop and a teahouse on the crossroad of Nanjing and Zhejiang roads," says Qian Zonghao, a Tongji University professor and author of the book *Nanjing Road (1840s-1950s)*.

Ma rented the site from a British landlord, established the Shanghai Sincere Co. and built a 7-floor classic Baroque building in three years. The name "Sincere" displayed his business idea of selling good-quality products at honest, transparent prices.

According to a report on *the North-China Herald* on October 20, 1917, the general construction of Sincere was a reinforced concrete skeleton filled with bricks and stones. In the center of the store was a large "light well" with an ornamental skylight.

There were two enclosed concrete staircases to the store. The external walls had blue Ningpo stones all the way to the first floor and Japanese granite chip cement above them. The columns, the cornices and dressings were of patent stone.

"The store is the largest and most up-to-date Chinese-owned store in Shanghai, at which everything from a needle to an elephant (toy) can be obtained. The ground floor has sections for grocery and provisions, wines, spirits, liquors, kitchenware, hardware, cigars, tobaccos, smoker's requisites, patent medicines, perfumes, confectionery, etc. The first floor is for drapery, millinery, tailoring, haberdashery, furs, etc. The second floor is for porcelain and other household supplies, electric goods, sports goods, toys, musical instruments, etc. On the third floor are the general offices of the company and the furniture department," the report said.

"The general external style of the premise was Western neo-classical with Baroque elements. It was a typical commercial architecture with a tower on the corner and ornamented columns," says professor Qian.

As the first Chinese-owned big department store, Sincere enjoyed considerable success. Ma and his

team tried out novel ideas for promotion, such as a roof garden and even published advertisements to hire women as shop assistants.

"Traditional Chinese women were supposed to be housewife, so Sincere's attempt to hire shop assistants was like 'the first one to eat a hairy crab.' No one responded to the ad for one month. Then, Ma's wife Ho Qingtang and her two sisters-in-law volunteered to be the first ones, which was a great advertisement for Sincere Co.," Zhang says.

According to Shen Ji's book *Old Shanghai Nanjing Road*, the department store customers were mainly Chinese. Outdone by the domestic stores both in size and in business scale, their foreign rivals took a big hit. The four Chinese-owned department stores became a symbol of Nanjing Road.

Today the Sincere Co. building is shared by Jinjiang Star Hotel and Shanghai Fashion Store. The latter sells a variety of Chinese cashmere sweaters and clothes for middle aged and the elder-

ly. Among the other three stores, Sun Sun is now the Shanghai No. 1 Food Store. Wing On Department Store and The Sun, which housed the Shanghai No. 1 Department Store, have both been renovated and strive hard to revive their past glory.

"The four stores not only kicked off a new era of Shanghai's modern department stores, but also guided the city's modern trend and altered Shanghainese habits and lifestyles. The grand buildings are still there and archives of the four stores are well preserved by the Shanghai Archives Bureau," Zhang Yaojun says.

Yesterday: The Sincere Co Building **Today:** Shanghai Fashion Store
Address: 299 Nanjing Road E. **Date of construction:** 1917
Architectural style: Western neo-classical with Baroque elements
Tips: Please note the change of architectural styles of the four big department stores on Nanjing Road while Sincere Co. Building, built in 1917, boasts a classic style.

著名建筑师唐玉恩观点
The Vision of Architect Tang Yu'en

南京路比外滩内容更多，有着跟外滩不一样的建筑。第一届20世纪中国建筑遗产评选时，我曾推荐了南京东路建筑群，以四大公司为主，还加了金门饭店和国际饭店。

"四大公司"的创办者把国外流行的百货大楼包罗万象的商业模式引进。"四大公司"的建成使南京路马上显出与其他道路的不同。

这些商业建筑不同于中国传统的商店。南京路早期的商店把中式屋顶升高做出牌坊的样子，而这"四大公司"是西式的，面积扩大了很多。中国的商店历来沿街沿河分布，都是线形的。而"四大公司"规模一下子扩大了，占了一个街坊，从此有了现代商业理念。进出货也不一样了，客人从两条大马路进出，货物从小马路进出，这四个建筑赢得很高的评价，使南京路的品质超过了其他马路。

外滩有江景而南京路没有，但这几家公司都做了很好的屋顶花园。夏天举行舞会和音乐会，提供餐饮服务，成为包揽市民各种消费的综合大楼。这几座建筑也反映了远东上海这座城市的特点。今天，豫园保留着中国传统的街道小商铺的商业模式，而南京路体现的是大城市现代化集中的综合功能。

The four emporiums introduced to Shanghai the business model of department stores which was in vogue overseas early last century. Their establishments made Nanjing Road stand out from other shopping streets.

The four stores are so different from traditional Chinese commercial architecture featuring Chinese-style big roofs and memorial gateways. Traditional Chinese shops which usually lined along a street or a river were arranged in a linear way. The four department stores have a much larger scale, covered a whole block and was built with an advanced modern commercial idea that products were transported through inner smaller streets while customers entered through two larger streets.

The four department stores had very nice roof gardens which hosted dancing parties or concerts in summer. They offered a variety of services and entertainments for customers. The four buildings dramatically up-graded Nanjing Road and reflected a Far Eastern city like Shanghai early last century.

从巴洛克到七重天

Wing On Towered over Its Rival

永安百货公司是南京路四大公司开业的第二家。与第一家先施公司相比，永安百货不仅生意更好，还建成当时南京路的第一高楼。

英文《上海泰晤士报》将永安公司的成功形容为"上海商业建筑里以建筑规模取胜的著名案例"。

永安大楼包括百货商店和大东饭店两个部分，担纲设计的公和洋行是20世纪二三十年代上海最主流的建筑事务所，外滩临江23座历史建筑中有9座都是公和洋行的作品。大楼主入口位于南京路上，在今天的浙江中路、九

江路和金华路上都有入口。

1918年8月26日永安开业前夕，《上海泰晤士报》报道形容由四条道路环绕的永安大楼所在地块，仿佛外国租界中央的一个小岛。

"建筑师的设计非常成功，立面优雅雄伟，设计有大型橱窗展示和广告空间，兼顾商业需求。这些空间由壁柱装点，没有影响整体的设计方案。"报道写道。

同济大学钱宗灏教授认为，永安大楼这座建筑比一年前开业的竞争对手先施公司大楼更加华丽。

"先施公司以西方新古典主义为主，巴洛克装饰为辅，而永安百货以巴洛克风格为主，新古典主义装饰为辅。永安公司的巴洛克装饰更加华丽，开业后就把先施比了下去，永安后来又投资建造了七重天摩天楼。"他说。

1918年的英文报道详细描述了这座巴洛克建筑的风貌。大楼主入口由一行双柱和外廊装饰，如今仍保持历史原貌。

"外廊地坪铺有马赛克，石膏天花板中央有方格，装饰有双柱和壁柱。建筑顶部有一个高耸的塔楼。展示橱窗是上海所有商店橱窗里最整洁和最摩登的。"报道写道。

为了在南京路激烈的竞争中胜出，1932年永安公司聘请美国建筑师哈沙德（Elliot Hazzard）在老楼边的三角形基地盖起一座19层高的摩天楼，与老楼以天桥相连。如今，这座简洁的钢构巨厦仍屹立在永安老楼旁边。

研究南京路"四大公司"档案的上海档案馆研究员张姚俊发现，永安公司的成功源于营销策略。1918年8月20日开始，创立

永安百货的郭氏兄弟便在《申报》上连续刊登广告。9月5日开业后，公司原本准备销售两个月的货品在20天内售罄，当时的盛况由此可见一斑。

"四大公司都绞尽脑汁用各种促销手段吸引顾客，而永安百货将女营业员用到了极致。他们招聘年轻貌美、端庄秀丽又略懂英文的上海小姐推销美国康克令金笔，销售非常'火爆'。1936年，著名新闻人徐铸成到沪后专程去永安买笔。他细细打量了女售货员，对举止优雅的'康克令小姐'发出了'果然明眸皓齿，不负众望'的赞叹。"他说。

此外，永安百货每年四季和周年庆都举办大促销活动，并提供类似信用卡的"记账消费"服务，让高端客人可以先购买后付款。永安百货还是上海最早设置

橱窗展示商品的商店。此外，室内时装秀、送货上门、发送礼券等服务也让商店增色不少。"永安"的好口彩让礼券成为深受欢迎的礼品。

永安公司和先施公司都有屋顶花园，为顾客提供了休闲空间。回到百年前的南京路，在这座巴洛克大楼里购物休闲，应该是绝佳的享受。英文报道提到，商场的地面铺着红色镶边的白色水磨石。试衣间和展示橱窗都由柚木打造。百货公司的一到四层销售全球精选的商品——从香水、牙膏、丝绸、珠宝到家具和行李箱，琳琅满目。

五楼设有顾客休息室，六楼是娱乐室和灯光茶室花园。屋顶花园包括夏屋和凉亭，有高耸的塔楼，晚上点灯后给人一种童话花园的感觉。而塔楼之巅曾是上海最高的地方，眺望城中风景的视野绝佳。

1949年5月上海解放，永安职员中的中共地下党员将一面五星红旗升起在永安顶楼的绮云阁，这是在南京路飘扬的第一面红旗。

此后，永安百货的命运也发生了变化，成为国营第十商店。1988年的整修对大楼的历史旧貌改动较大，商店重新开张后改名华联商厦。2004年，巴洛克大楼又历经修缮，清除了玻璃幕墙，已被水泥覆盖的铸铁栏杆也被剥离出来。2005年工程竣工，商厦恢复了1918年的历史旧貌和好口彩名字——永安百货。

2018年9月5日，永安百货隆重举办了100周年庆典活动。

昨天：永安百货　**今天**：永安百货　**地址**：南京东路 635 号　**建造时间**：1918 年
建筑师：公和洋行　**建筑风格**：巴洛克风格
参观指南：大楼内三楼和四楼是一家中餐厅，保存了大量室内历史细节，值得欣赏。

The Wing On Department Store was the second of the four big Chinese department stores on Nanjing Road — the other three being Sincere Co., Sun Sun and The Sun. The emporium not only enjoyed better business than the first one, Sincere Co., but also built the tallest building on the popular shopping street.

The success was described as "a notable example of shop building in Shanghai on an architectural scale" by *the Shanghai Times* in 1918.

The building, consisting of a spacious department store and the Great Eastern Hotel, was designed by Palmer & Turner, a leading architectural firm which had designed nine of the 23 waterfront buildings on the Bund.

The Wing On building had its main entrance on Nanjing Road with on today's Zhejiang, Jiujiang and Jinhua roads.

"The whole block, embraced by the four roads, forms an island site in the very center of the traffic of the foreign settlement. The architects have achieved a complete success, the elevations being both graceful and imposing, while at the same time complying with the commercial requirements for a large window display and spaces for advertising. These spaces have been allowed for the pilasters without in any way detracting from the general scheme of design," *the Shanghai Times* reported on August 26, 1918, shortly before the store's opening.

Tongji University professor Qian Zonghao says Wing On had a more elaborate appearance than its competitor, the Sincere Co., which opened a year earlier.

"Sincere Co. was designed in neoclassical style with Baroque elements as supplement. Wing On, on the other hand, had Baroque style with some neoclassical adornments. Thus, Wing On's Baroque decor was more appealing and exquisite and attracted customers. After Wing On opened in 1918, Sincere could not rival it," says Qian.

The report on *Shanghai Sunday Times* described the building in detail, which still had the main entrance marked by a series of coupled columns that formed a covered arcade.

"The floor of the arcade is mosaic and the ceiling of modeled plaster covered with coffered center. The coupled columns and central feature are superposed with pilasters and crowned with a lofty tower. The show windows are of molded copper. All the sun blinds are covered and boxed in the most neat and modern group of windows for displaying articles in Shanghai," the newspaper reported.

To win from fierce competition on Nanjing Road, Wing On erected a new 19-floor building in 1932 on the triangle-shaped site adjacent to the 1918 premise. Designed by American architect Elliot Hazzard, the new building was a simply-cut, steel-structured skyscraper, which still stands next to the century-old Wing On Department Store. In addition, Wing On employed young, beautiful, English-speaking girls as saleswomen for the American Conklin fountain pens, which was a very successful promotion.

Shanghai Archives Bureau researcher Zhang Yaojun also attributes Wing On's success to the

store's varied smart promotions. From August 20, 1918, the Guo brothers, founders of Wing On, regularly advertised the upcoming launch of the store in local Chinese newspapers. Within 20 days since its opening on September 5, its stock for two months was sold out.

"All the four department stores racked their brains and worked out a rainbow of creative promotions to attract customers. It was Wing On that made full use of women shop assistants. They employed young, beautiful English-speaking girls as saleswomen for the American Conklin fountain pens, which proved to be very successful. Visitors to Shang-hai, including famous journalist Xu Zhucheng, would visit Wing On to buy fountain pens while admiring the charming 'Miss Conklin'," Zhang says.

Wing On was operated in a flexible, innovative manner, hosting five big sales in the four seasons and one anniversary sale every year. The shop also created "accounting consumption" , similar to credit card today, for upper-class customers to buy and pay later. Wing On also invented shop window display, organized indoor fashion shows, offered delivery service and issued gift coupons to attract customers. The meaning of Wing On in Chinese, "eternal safety," made its voucher

coupons a popular gift source.

Both Sincere Co. and Wing On had roof gardens and offered a variety of entertainment options besides shopping. It might have been a great experience to shop and relax in this Baroque building on Nanjing Road. Archives showed that the floors were made of white terrazzo with red borders, while the shop fittings and display cases were made of teak. The ground four floors sold a galaxy of globally selected products — from perfume, tooth paste, silk, jewelry to furniture and suitcases.

The fourth floor was partly used as a refreshment and rest room for customers. The fifth floor had amusement rooms and tea garden illuminated with lights. The roof garden comprising summer houses and arbors had a crowning feature in the lofty tower. The entire place, when lit up, produced an effect of a fairy garden. The summit of the tower, which was the loftiest in Shanghai, offered the most extensive views of the city.

When Shanghai was liberated

in May 1949, the first red flag was raised on top of Wing On building. Wing On later became the state-run Shanghai No. 10 Department Store.

In 1988, the store reopened as Shanghai Hualian Department Store after renovation with many historical details changed. During another restoration of the Baroque building in 2004, glass walls added in 1987 were removed and old cast iron railings, already wrapped in cement railings, were redone. When the project was completed in 2005, the emporium was renamed Yongan Department Store, almost similar to its old name (the same in Chinese but different in pinyin) and the old look of 1918.

On September 5, 2018, the store celebrated its centenary with a grand ceremony.

Yesterday: Wing On Department Store   **Today:** Yongan Department Store
Address: 635 Nanjing Rd E.   **Date of construction:** 1918
Architect: Palmer & Turner   **Architectural style:** Baroque
Tips: The third and fourth floors, which house a Chinese restaurant, have retained historical details and are worth a visit.

永安新厦"七重天"
The Seventh-Heaven Annex

1932年，永安公司位于湖北路浙江路口的19层摩天新厦落成，上海多家英文报纸纷纷报道。英文《大陆报》提到，新楼有五层的空间为永安公司增加了营业面积。第六层和屋顶部分用作娱乐空间，设计有电影院、茶室花园和室外漫步空间。新楼与老楼由一座天桥相连，建成后成为远东最高的建筑之一，永安公司高管的办公室也会设在这里。

"建筑外立面设计为现代风格，以垂直线条为主。底层直到二层窗台都是苏州花岗岩。每一面都有大玻璃橱窗。厚玻璃橱窗底部是绿色大理石基座，有石砌装饰。垂直的壁柱可能用浅色人造石或仿石石膏制成，上部有金属拱肩，下面是钢窗扇。摩天楼顶部会有有趣的照明系统，将为上海城市的天际线增添一枚新的更高的音符。"1932年6月9日的《大陆报》报道写道。

今天，这座南京路摩天楼里有一家"七重天宾馆"，酒店名字源于当年楼里的"七重天餐厅"。

在20世纪30年代，在19楼的七重天之巅，人们可以眺望全城的景色，包括老城区、闸北、浦东和黄浦江。

2017年，我有幸在大楼业主黄浦置业工作人员的陪同下，攀爬一条近似垂直的无扶手钢梯，钻过一个十分窄小的洞口后，抵达七重天之巅。今天在此眺望全城，虽然不像当年一样能看到老城区和闸北，但仍然可以欣赏几乎整条南京东路的景色。熙熙攘攘的人流两旁，繁华林立的商厦大楼间，一座座历史建筑点缀其中，静默地诉说百年前的芳华。

The 19-floor annex of the Wing On Department Store that was constructed at the intersection of Hubei and Zhejiang roads was widely covered by the city's English newspapers in 1932.

A report in the China Press said the new building provided five floors of additional space for general sales. The fifth floor and the extensive roof spaces were intended for amusement purposes, including a cinema, a tea garden and an open-air promenade space.

An important feature of the new store was a bridge that connected the store with the annex. The tower at the Nanjing Road intersection, containing the offices of managing directors, was one of the highest structures in the Far East.

"The general style of architecture proposed for the exterior was the modern style with vertical lines predominating. The lower story is of Soochow granite up to a belt course at the sills of the first floor windows. Large shop windows are provided on all fronts with green marble bases below the plate glass and between the masonry pieces. The vertical piers will probably be of light colored cast stone or imitation stone plaster with moulded spandrils of metal above and below the steel casement sash. In the upper part of the tower will be provision for an interesting system of lighting, which is intended to add a new and higher note to the Shanghai skyline," *the China Press* said in a June 9, 1932 report.

Today the building houses the state-run Seventh Heaven Hotel. The name was derived from a former renowned restaurant in the annex.

Back in the 1930s, the observation gallery on the 19th floor allowed visitors to see the entire city including the former old town, today's Zhabei District (now part of Jing'an District) and the Pudong New Area and even the wide stretch of the Huangpu River on clear days.

It's a perfect spot (but unfortunately not open to the public) to admire the entire stretch of Nanjing Road E. all the way to the Bund, graced by a galaxy of historical buildings on both sides.

新新公司的新意

Sun Sun Co. profited from novel ideas

南京路"四大公司"第三家开业的新新公司，名字有深意。

新新之名源于《大学》中的一句话——苟日新，日日新，又日新"。1926年新新公司开业，比前两家南京路百货公司晚了近10年，当时南京路的地价已涨了不少，新店要有更新奇的营销策略才能在激烈竞争中获胜。

新新公司坐落于南京路广西路口，由曾在澳洲经商的刘锡基创办。刘锡基回国后曾在南京路"四大公司"中最先开业的先施公司担任经理。1924年他辞职创业，只用了不到两年的时间，就与其他商人一起开办了新新公司。

1926年1月，新新公司开门试营业，英文北华捷报的报道形容"这里商店宽敞大气，与众不同"。

商店一楼陈列葡萄酒和烈酒、办公用品，此外还有一家储蓄银行。二楼销售丝绸和鞋靴，三楼是瓷器、玻璃制品、电器、钟表，四楼供应铜铁床架、地毯和红木家具。开业当天，新新公司吸引了超过5万名中外顾客，

他们发现"在芝加哥能买到的东西在这里都有，有些在号称'百货公司之都'的芝加哥都买不到的商品这里也有。"

报道评价，新新公司的新厦"体现了上海的发展与未来的繁荣"。"新大楼设计有最现代的线条，窗户很多，照明都是非直射的，还有舒适的电梯和充足的展示柜台，会令满怀期望而来的顾客高兴，让购物成为一件愉悦的事。"《北华捷报》报道。

同济大学钱宗灏教授认为新新公司的建筑风格比前两家南京路百货公司要更加现代。"外立面以三段式划分，体现了装饰艺术风格，或者更准确地说，是古典主义向装饰艺术过渡的风格。"钱教授说。

设计大楼的是匈牙利建筑师鸿达。这位一战后从西伯利亚战俘营来到上海的建筑师擅长摩登现代的风格，其代表作还包括淮海路国泰电影院、外滩14号总工会大楼和位于外滩源的光陆大楼。在南京路上，鸿达还操刀过1930年惠罗公司的改造工程，将这座19世纪古典主义风格的老牌

百货公司改造成一家简洁现代的商店。

"大楼马赛克装饰和坚固栏杆展现出现代主义气息，而背立面金属栏杆的装饰则具有装饰艺术派风格的多重特征，体现出新艺术运动时期分离派的最后一丝精神余脉。建筑外立面形式与美国建筑设计先驱路易斯·沙利文创办的芝加哥学派一脉相承。"匈牙利出版的《鸿达》一书如此评价这件作品。

同济大学卢永毅教授也认为新新公司的立面构图与芝加哥学派框架结构的建筑相近。"而顶部超常的标志塔楼，将其塑造成一座风格清新的商业纪念碑。"卢教授在论文《近代上海四大公司公司》中写道。

她提到，百货商厦的建筑设计也参与到大都市空间和资本的竞夺之中。"建筑高度的争夺一方面反映四大公司既要盘踞一团又要相互竞争的局面，但更重要的另一方面，是要在华洋杂居的租界中，争得民族资本的地位。"

从1926年开业那一天，新新公司名副其实地不断大胆创新。根据《北华捷报》报道，公司在一楼设置了香水球，香水洒落在楼梯上，清香四溢。顾客们纷纷用手帕或衣领接香水滴。《大陆报》还提到，最让人眼前一亮的是商店宽大的中央楼梯，两侧还有可以容纳20人的电梯。此外，新新公司的营销创意还包括赠送免费香烟、邀请"上海最漂亮的女孩子"展示现代泳装和举办玩具展等。

不过，研究"四大公司"档案的上海档案馆研究员张姚俊认为，新新公司最出彩的创意还是玻璃电台。这是因为刘锡基不想模仿先施公司和永安百货的屋顶游乐场，于是想出玻璃电台的点子。

当时国内第一家电台开播不久，电台节目是相当新鲜时髦的，大多数电台都是外资背景。曾在美国学习无线电技术的新新公司技师研制了电台设备。1927年3月18日，新新电台在商店六楼开播。

电台每天播音6小时，既有戏曲节目，又有店内促销和产品

介绍。后来，上海的电台越来越多，刘锡基就把播音室做成全透明的玻璃屋，让人们能看到原本神秘的播音过程。从此，"玻璃电台"成为新新公司的一大卖点，对商店的营业额也有不少提升。

如今，"玻璃电台"已成为历史，摩登大楼成为上海第一食品商店南京路旗舰店。商店四层楼面琳琅满目地陈列着2万种中外美食，来自全国各地的顾客忙碌挑选着，生意火爆，年营业额达6亿元。难以想象在同样的空间里，曾有年轻女郎坐在玻璃房里，用温柔的软语播音。

著名音乐家、《梁祝》作曲陈钢先生的母亲金娇丽就在此播过音。"1926年1月23日建成开业的上海'新新公司'6层的新都饭店，别开生面地在大厅里自行设计，自行装备了第一个由中国人创办的广播电台，因电台的房子四周是用玻璃隔断的，故俗称'玻璃电台'。这也许可以视为当年电台透明、开放的某种象征，和白领女性走进都市生活的一个标志。我母亲当年就曾在'玻璃电台'里担任过播音员。"作曲家在一本关于上海老歌的书里写道。

这本书的名字，就叫《玻璃电台》。

昨天：新新公司　**今天：**上海第一食品商店　**地址：**南京东路 720 号
设计师：鸿达　**建筑风格：**装饰艺术风格
参观指南：先欣赏这座建筑沐浴阳光的简洁立面，再走进这家食品商店体验琳琅满目的传统上海美食。

The people behind Sun Sun Co., the third of the four big Chinese department stores on Nanjing Road (the other three being Sincere Co., Wing On and The Sun), borrowed a bit from ancient Chinese classics to mark itself out from its competitors and strike a chord among the locals.

The names, Sun Sun and The Sun, both contained the Chinese character "xin" or "new". Inspired by a line from the Confucian classic "Da Xue", or "Great Learning", the store names implied making new progress and achievement every day.

Unlike the first two stores-- Sincere and Wing On--Sun Sun Co and The Sun had to dig deep into their pockets to rent a place on Nanjing Road because of soaring land price. Despite that, the two stores fearlessly marched into this popular street, and strived to make innovations to surpass their competitors.

Founded by Chinese merchant S. K. Lau, Sun Sun Co. opened in 1926 on the crossroad of today's Nanjing and Guangxi roads. With experience of business in Australia, Lau returned to China and joined the Sincere Co. of Shanghai as a sub-manager and was later promoted as a manager. He resigned in 1924 and set up the Sun Sun Co. in collaboration with a group of enterprising businessmen within a short period of a little more than two years.

When the emporium's doors were "thrown open for inspection" in January 1926, it was described as a "spacious, commodious and different department store" by *the North-China Herald.*

The ground floor showcased wines and spirits, stationery and a savings bank while the first floor contained silks, boots and shoes. The second floor was devoted to chinaware, glass, electric fixtures, clocks and watches while the third floor was stocked with brass and iron bedsteads, carpets and black wood furniture.

More than 50,000 foreign and Chinese patrons went from floor to floor on the day of its opening and "found almost anything that one can 'buy in Chicago,' and perhaps even some articles that are not purchasable in the metropolis of department stores."

The edifice was also "a tribute to Shanghai's growth and future prosperity." Tongji University professor Qian Zonghao says the facade divided into three sections displays Art Deco style, or more accurately a transition from neoclassic to Art Deco style.

"Constructed on the most modern lines, plenty of windows and good indirect lighting, comfortable lifts and adequate display counters, the new premises will delight prospective purchasers and make shopping a pleasure," *the North-China Herald* said in a report.

The architect was C. H. Gonda, a renowned Hungarian architect, whose works included the Cathay Theater on Huaihai Road, No. 14 on the Bund and the Capitol Theater in today's Waitanyuan, all of which feature an ultra-modern style. In 1930, the architect also renovated the 19th-century building of Whiteaway, Laidlaw & Co. Ltd. on Nanjing Road, which at one time was the largest foreign-owned department store in modern Shanghai.

Lu Yongyi, another Tongji University professor in architectural history, notes the facade of Sun Sun Co. mirrored the Chicago school of architecture.

"Sun Sun's extraordinary signature tower fashioned the building into a commercial monument of a fresh style," she wrote in a paper titled "The Four Department Stores of Shanghai in the Early 20th Century".

In the era of the four department stores, architectural design participated in the competition for space and capital in metropolitan Shanghai, Lu notes.

"The strenuous efforts to achieve architectural heights reflected congregation and competition of the four stores, but more importantly, the position of Chinese capitals in a foreign settlement," she says.

Sun Sun Co. lived up to its name and boldly experimented with innovations since its opening in 1926. *The North-China Herald* mentioned that "one of the most popular devices installed in the new store of the Sun Sun Co. is the Perfume Ball from which drips scented spray down on the first step of the stairway on the ground floor." It attracted customers who waited to catch every drop on to their handkerchiefs or coat sleeves.

The China Press noted that what "strikes the eye at once is the spacious central stairway, flanked by roomy lifts capable of carrying twenty persons."

In addition, Sun Sun Co. also worked out a variety of promotional ideas such as giving out free cigarettes, inviting "some of the city's prettiest girls" to parade for an exhibition of modern swimsuits or hosted toy shows.

But the store's greatest promotional idea was a "glass radio station". Back in the 1920s, radio programs were still new to many Chinese as the first wireless radio station opened only in 1923 and the stations were all owned by foreign companies.

A Lau's employee, a technician named Kuang Zan who had studied wireless technology in San Francisco, agreed to design and manufacture the equipment for Sun Sun's radio station. It began to broadcast pleasant music from March 18, 1927 from the store's sixth floor.

The station broadcast six hours every day featuring mostly traditional opera programs while promoting the store's products and special promotions.

"When more radio stations opened in Shanghai, smart Lau renovated the broadcasting room into an all transparent glass room. The 'glass radio station' allowed people to see the usually mysterious broadcasting process. It became a major attraction and helped in increasing the store's sales considerably," says Zhang Yaojun, a researcher from Shanghai Archives Bureau.

Today, the glass radio stations are a thing of the past but the Art Deco building is still blessed with good business by serving as Shanghai First Food Hall, a Nanjing Road flagship of Shanghai First Food Chain Development Co. Ltd..

The 10,000-square-meter food hall offers more than 20,000 kinds of food products in its four floors and boasts an amazing sale of 600 million yuan (US$92 million) every year. A paradise for food lovers, the mall is filled traditional Chinese or imported food and flooded with customers from around the country. You can now only imagine former scenes in the building when a beautiful young lady sat behind a glass and in a soft-toned voice spoke into the microphone.

Renowned Shanghai composer Chen Gang's mother Jin Jiaoli was a broadcaster at the Sun Sun radio station.

"It was the first radio station to be designed, equipped and founded by a Chinese. I think it was also a kind of symbol of transparent, open radio station and white-collar Chinese women began entering metropolitan life," the 83-year-old musician wrote in a book themed on old Shanghai songs and singers. He titled the book *the Glass Radio Station*.

Yesterday: Sun Sun Co.   **Today:** Shanghai First Food Hall   **Address:** 720 Nanjing Rd. E.
Date of construction: 1926   **Architect:** C. H. Gonda   **Architectural style:** Art Deco
Tips: Admire the building's simple-cut Art Deco facade bathed in sunlight before entering the food hall which offers one of the city's most complete collections of traditional Shanghai food.

新新公司的圣诞节一站式购物
One-stop Christmas shopping

南京路"环球百货"新新公司的柜台对于圣诞节购物者来说非常诱人……

该商店宣称过去一周的圣诞节销售数额很大。最近为了吸引选购圣诞节礼物的顾客，商店进口了许多优质玩具和游戏。

这家现代化的百货商店让圣诞节购物变成一件轻松的事情。在新新公司购物很简单，因为所有商品都陈列在相应的部门。此外，今年商品的价格异常低廉，这一点也很吸引人。

在节日期间，整个商店装饰得喜气洋洋。您要做的就是列一份圣诞节购物清单，然后前往新新公司，让他们为您完成订单。

新新公司的玩具部门货品丰富，有各种各样的玩具，包括大大小小的娃娃、栩栩如生的动物玩具、机械套装玩具、数百种适合男孩和女孩的游戏和精美书籍。

这些展示的礼物都很"完美"。圣诞期间会出售精美文具、钢笔和其他书写工具，价格非常适中。

新新公司还有大量精致的新奇珠宝、香水和许多小物件，它们会给圣诞带来惊喜。

在食品部，可以购买好看的盒装巧克力、装在圣诞袜里的糖果、饼干、甜食、优质烈酒和葡萄酒。

在新新公司为全家人采购非常容易。对于吸烟者来说，这里有齐全的烟斗、烟嘴、高档雪茄和香烟。对摄影爱好者来说这里有优质的相机销售，也有分别适合老年人和年轻人穿着的服饰。

如果您打算为今年的圣诞节购买一条领带，一定要看看新新公司的领带，那里既有适合保守人士的安静风格的领带，对于喜欢一点色彩的人也有装饰华丽的款式。

无论您想买什么东西，在新新公司基本都可以找到。领结、服饰用品、相机、香水或圣诞节大餐的食材，您都可以在新新公司找到。

Sun Sun Company, Shanghai's Latest Department Store, Will Open Tomorrow

Nanjing Road is thronged with shoppers from all over China, especially during the holiday season like Christmas, New Year and the Chinese New Year.

Back in 1936, *the China Press* published a special report on wide choices of gifts on offer at Sun Sun Co..The story reflects the living standards and lifestyles of the well-off in old Shanghai:

"The counters of the Sun Sun Department Store, the Universal Providers, on Nanking Road, would tempt any Christmas shopper …

The store reported a great volume of Christmas trade in the past week. The large stock of excellent toys and games, recently imported by the store, is proving to be a big draw card for Christmas present seekers.

Everything to make Christmas shopping easy has been done at this modern department store. It is easy to shop at Sun Sun because everything is placed in its proper department. Another attraction is that prices are unusually low this year.

The entire store is gaily decorated for the holiday season. All you have to do is to make a Christmas shopping list and go to Sun Sun and have

them fill the order for you.

The toy department at Sun Sun is very complete, with a wide variety of toys being displayed. Dolls, small and big, animated toy animals and mechanical sets, hundreds of suitable games and attractive books for boys and girls are among the many things you can choose from.

From top to bottom the gifts displayed are 'perfect'. Fine stationery, fountain pens and other writing accessories are on sale for Christmas, and the prices are very moderate.

Sun Sun has a great stock of unusually fine novelty jewelry, perfumes and any number of little things that are bound to make pleasant Christmas surprises.

In the provision department, fancy boxes of chocolates, candy in stockings, biscuits and confectionary, fine spirits and wines can be bought.

It is easy to do your shopping for the entire family at Sun Sun Department Store. The collection of pipes, cigar and cigarette holders, fine cigars and cigarettes and pipe tobacco would tempt any smoker. There are excellent cameras for those who have a hobby in photography, and wearing apparel to suit the needs of young and old.

If you plan to give Christmas ties this year be sure to look over Sun Sun's stock of ties. You'll find them in quiet styles for the conservative man and in elaborately decorated styles to suit the man who likes a bit of color.

Whatever it is that you want to purchase the chances are 99 to one than you'll find it at Sun Sun. Neckties and other wearing apparel things for the smoker, cameras, perfumes or provisions for Christmas dinner — you'll find them all at the Sun Sun Department Store.

摘自 1936 年 12 月 22 日《大陆报》
Excerpts from *the China Press*, on December 22, 1936

南京路的"中国橱窗"

The Sun: the show window of China

大新公司1936年才开幕，是南京路"四大公司"里最晚开业的一家。这家百货公司名副其实，与前三家店相比，规模最大，设备最新。2018年11月13日，位于昔日大新百货的一百旧楼和两座新楼改造合并为第一百货商业中心，在漫天金粉中隆重开业，开启了新一段"大新"的历史。

大新百货的创始人蔡昌（Choy Chong）与南京路其他三家百货公司的创始人背景类似，曾在澳洲经商，在香港和广州创办了大新公司。公司位于南京路的大楼高达44.5米，营业面积近30,000平方米，三面临街都有入口。1935年10月大楼即将竣工时，英文《大陆报》称这座百货公司新厦被打造成了"中国的橱窗"。

"商场里有很多创新。也许大众看来最出色的就是商场使用了自动扶梯。大新公司声称该这是第一次在中国引进两部自动扶梯。"报道提到。

1936年大新公司开业，两部奥的斯自动扶梯吸引好奇的人们蜂拥而至，公司不得不售卖乘坐电梯票控制人数。开业当日来店顾客有4万多人，自动扶梯成为最好的广告。

除了自动扶梯，商场利用地下一层营业是又一创新之举。大新公司的营业区域包括地下一层和地上三层，行政办公用房在四楼，其余楼层设有饭店、剧院、娱乐公园和屋顶茶馆，非常类似今天的购物中心。人们购物之余还能享受美食和娱乐。

同济大学钱宗灏教授认为，南京路四大公司建筑反映了上海近代建筑风格的变迁。建于1914年的先施公司以西方新古典主义为主，巴洛克风格为辅，而4年后更华丽的永安百货以巴洛克风格为主，新古典主义为辅。1926年，新新公司体现了新古典主义向装饰艺术风格的过渡。作为南京路四大公司中最年轻的一员，大新公司呈现了现代派风格，局部带有装饰艺术风格和中国民族风格的装饰。"大楼有现代派建筑的流畅，立面构图竖向线条为主。顶楼的装饰有一排中国挂落，还有回字纹和几何纹等中国

式的装饰手法。"钱教授谈道。

设计这座国风摩登商厦的是基泰工程司，由关颂声和杨廷宝等几位美国宾夕法尼亚大学留学归国的近代中国建筑师创立。基泰工程司在全国多个城市都有作品，在上海设计了中山医院和九江路大陆银行大楼等。

1935年的英文报道提到，大新公司业主选择了这种简洁的设计风格，而牺牲了过多装饰，以便"为顾客提供更好的服务"。公司的经营特色是贴心服务，营业员几乎都会讲英语。大新公司还提供信息服务，可以告诉顾客在本店或者其他哪家商店能买到所需的商品。以方巾销售为例，经营团队为满足游客需求，为方巾类产品专门设立了展销部门。

"中国艺术风格的方巾花色繁多，全世界没有哪个国家可与之相比。往来上海的游客对于方巾艺术常设展都兴趣浓厚。上海是远东最大的大都市，拥有许多亚洲最高的建筑。"报道写道。

1951年，大新公司停止营业，国营上海市第一百货商店于1953年迁入营业。在20世纪八九十年代，第一百货也如昔日大新百货一样名副其实，曾经蝉联十多年全国商场销售冠军。1993年第一百货股票上市，1997年一百东楼开业，通过三层玻璃天桥与老楼连为一体，2007年一百新楼建成，2017年一百闭店修缮改造。

2018年11月13日，完成改造的一百商业中心隆重开业。商业中心由一百老楼、新楼和东楼组成，通过多条跃层飞梯和空中连廊互通连接。新老大楼间原有三条连廊都已老旧，这次改造除翻新了老连廊，还新建了位于三层、五层、七层的四根连廊，长度约14至20米。老楼作为文物保护建筑外立面不允许变动，因此新增的四根连廊中，有三根连接一百新楼和东楼，只有一根连接到老楼顶部七层。

除了连廊以外，新老大楼之间还加盖了一个高达42米，长80米的透明顶棚。顶棚使用聚碳酸酯材质的透明管和实心板，仿佛为六合路撑起了一把巨大的透明伞，为七条连廊遮风挡雨。

大楼之间的六合路作为南京

路步行街的后街，曾经嘈杂凌乱，这次也由透明的超大顶棚覆盖，重新布置绿化与服务设施后变身为新的城市休闲空间。2019年8月7日七夕夜，六合路举行为期五天的"夜游会"，在彩灯映照下开出一家家销售文创产品的帐篷店铺，成为南京路步行街上"夜经济"的新尝试。

第一百货改造项目历时497天，耗资40多亿元人民币，将营业面积由大新公司时代的3万平方米扩展到近12万平方米，是南京路城市更新的一个高潮。

一百商业中心的原副总经理庄倩提到，位于一座历史建筑里的第一百货面对现代购物中心的激烈竞争，希望通过这次更新改造吸引年轻消费者。除了空间改造，一百还加强了梧桐、弄堂等老上海元素，在7楼专设 "100弄"文化展览空间和"100里"

海派伴手礼体验店。 在这座1936年的百货公司里，到访上海的游客可以买到国际饭店的蝴蝶酥、第一铅笔的中华牌铅笔和上海制皂的蜂花檀香皂等上海特色礼品。

重新开业的第一百货在店堂内张贴了彩色海报，写着"老地方，new joys（新欢乐）"。为了让这座1936年的建筑在南京路的新时代重获"大而新"的美名，许多人为此付出了智慧和心力。

昨天: 大新公司 **今天:** 第一百货商业中心 **地址:** 南京东路830号
建造年代: 1936年 **建筑师:** 基泰工程司
参观指南: 建议参观位于商场七楼的"100弄"文化展览空间和"100里"海派伴手礼体验店。

The Sun, which opened in 1936, was the last of the four big Chinese-owned department stores on Nanjing Road--the other three being Sincere Co., Wing On and Sun Sun. It lived up to its Chinese name, Daxin, or "big and new". The store was fairly big in terms of size and boasted the most modern facilities among all the four stores.

The building, now housing the Shanghai No. 1 Department Store, reopened after a half-year renovation, which is part of a bigger redevelopment plan to turn the area into a classy Shanghai No. 1 Shopping Center with all the old Shanghai elements.

It was owned by Chinese merchant Choy Chong, who had business experience in Australia and set up the parent of The Sun Company in Hong Kong and Guangzhou.

The 44.5-meter-high store featured a shopping area of nearly 30,000 square meters and had entrances on three roads. When the nine-story building was nearly completed in 1935, *the China Press* reported that the founders had christened the store as "The Show Window of China".

"A number of innovations will be introduced. Outstanding, perhaps from the viewpoint of the public, will be the introduction of escalators. The Sun Company thus claims the honor of having introduced the first two passenger escalators in China," the report noted.

The two Otis escalators, imported from the United States, were so novel in 1936 that many locals thronged to the store simply to experience it. The Sun started selling tickets for a ride which would be deducted from any purchase. It was estimated that over 40,000 people visited The Sun on the opening day and the escalators became the best advertisement that ensured good business.

The company's shopping floor in the basement was another innovation --the first of its kind in China. In the 1930s, the basement and three floors including the ground level served as shopping areas. The third floor was used as Taipan's office and other administration purposes. The upper floors had a restaurant, a theater, an amusement park and a roof-garden tea house, pretty much like today's shopping centers where customers could not only buy products, but also enjoy food, beverage as well as entertainment.

Tongji University professor Qian Zonghao says the four big department stores mirror the evolution of Shanghai's modern architectural styles.

"The style changes from Sincere Co's neoclassical style with Baroque elements in 1914 to Wing On's Baroque style with neoclassical elements in 1918, and then switched to Art Deco with Sun Sun in 1926," according to Qian.

As the youngest of the lot, The Sun was designed in modern style with Art Deco and Chinese ornaments. "The simple-cut facade is graced by vertical lines and yellow glazed tiles," says Qian, adding that similar transition can also be seen on the Bund. "The Sun is a punctuation of the transition of Shanghai modern architectural styles on Nanjing Road."

The architect of the building was a big Chinese firm Messrs Kwan, Chu and Yang, which also designed the Continental Bank at 111 Jiujiang Road (now the office of Shanghai Trust) in Art Deco style. The contractors, the Voh Kee Construction Company, were behind the Park Hotel.

The 1935 report noted that the owners had chosen a simple design and sacrificed the ornate in order to "develop the best in service for the buying public".

The management team set aside special sections of the store for display and sale of handicraft commodities to meet the demand of the ever-increasing number of tourists because "no country in

the world produces a greater variety of artistic handicraft than China".

"This permanent exhibit of handicraft art objects will be of special interest to all travelers coming to or passing through Shanghai, the most cosmopolitan city of the Far East, and which boasts of having the tallest buildings in Asia," the report said.

The Sun prided itself in "community service". The majority of the salesmen and saleswomen behind the counters spoke English. For newcomers to the city, the information concerning both the merchandise sold in the store and objects of interest available elsewhere in Shanghai was offered.

A glass dome and a gallery are under construction to link the Shanghai No. 1 Shopping Center with adjacent structures.

In 1951, The Sun served its time and stopped business activity. Another state-owned emporium, Shanghai No. 1 Department Store, moved into the building in 1953.

The store lived up to its "No. 1" tag. In the 1980s and 1990s, it earned fame as "the largest shopping mall" and made top-selling records among all Chinese department stores for more than a decade. In 1993, the No. 1 store was listed on the Shanghai Stock Exchange. In 2008, the store grew further by building an adjacent new building.

In June 2017, the No. 1 store closed its door for the first time for redevelopment. The highlight of the refurnishing plan is a glass dome and a gallery in the air which connects with the Sun Sun Co., the annex building and the adjacent Oriental Department Store. The idea is to turn the traditional No. 1 department store to a modern 120,000-square-meter shopping center comprising three old and new buildings. Cultural exhibition space, a Shanghai-style souvenir store and stylish cafes opened up to bring in younger customers.

"With strong competition from modern shopping malls, we want to win more young consumers through this redevelopment that will add old Shanghai elements into the interior design, such as

alleyway houses and plane trees," says Zhuang Qian, former deputy CEO of Shanghai No. 1 Shopping Center.

When it reopened, colorful posts hype the new motto: "Same place, new joys." Efforts are being made to turn the 1936 building again into a "big and new" store in a new era on Nanjing Road.

Yesterday: The Sun Department Store   **Today:** Shanghai No. 1 Shopping Center
Address: 830 Nanjing Rd E.   **Date of construction:** 1936
Architect: Messrs Kwan, Chu and Yang
Tips: The store opens an exhibition center and a souvenir shop both themed on Shanghai culture on the seventh floor.

跑马总会的新时代

A New Era for Shanghai Race Club building

1933年的一个春日，上海跑马总会董事长伯克尔先生在来宾见证下，以夫人的名义为总会新楼和新看台安放奠基石。

"1933年5月21日，奠基石由凯瑟琳·伯克尔夫人安放，以纪念新楼的崛起，也纪念1863年的老看台被拆除。"奠基石上写着。

今天，这块奠基石依然镶嵌在上海历史博物馆北立面的左下角。2018年3月，昔日跑马总会俱乐部历经多年修缮，改造为上海历史博物馆对公众开发。博物馆从11万件馆藏里精了1000件文物展示，但主持修缮的著名建筑师唐玉恩认为，"上海历史博物馆最大的展品，就是这座建筑本身"。她认为，这座总会建筑蕴含了丰富的历史。20世纪50年代开始，它历经变迁，先后用作上海图书馆、上海博物馆和上海美术馆，建筑本身就映射了上海的历史。

1934年大楼竣工时，英文《大陆报》称这是"远东地区最好最奢华的的俱乐部"。

唐玉恩研究发现，大楼是一座典型的总会建筑。总会建筑是为满足租界外籍人士聚会社交的需求而建的一种特殊建筑类型，也是西方文化和城市生活传播到远东的体现。在老上海，各国都建造了富有本国特色的总会建筑，如外滩的英国上海总会和德国总会、福州路的美国花旗总会，还有两家法国总会——分别是今天的科学会堂和花园饭店。总会建筑的设计与商业建筑不同，倾向于使用轻松的乡村风格或类似该国民间建筑的形式。

"上海中心城区有数量众多的总会建筑，这在我国其他城市是不多见的。这些总会以会员制为主，提供酒吧、餐厅、阅览室、滚球房等设施，空间非常丰富，功能多样，是高端的交际场所。"唐玉恩介绍。

而上海跑马总会的功能与其他总会建筑也很相似。英文《大陆报》报道介绍，一楼有为会员和客人计算赌金的设备，二楼有休闲室、阅览室、会员咖啡屋、客人休息室和纸牌屋。楼上还有两座羽毛球场、两条英式和四条美式保龄球道、几个壁球场和公

寓房。看台可以容纳大量观众。

如今，昔日跑马总会俱乐部被称为"东楼"，以区别于历史博物馆院落中的"西楼"——一座建于20世纪20年代的马厩。在这次修缮中，"马厩"也被改造为博物馆的展陈空间。

"俱乐部"与"马厩"虽然建于不同年代，比例都很优美，反映了当时正流行的英国新古典主义建筑风格。与华丽的外滩银行建筑相比，跑马总会的两座楼很简约。

"建筑师并没有用大量石材，而是使用红砖与水刷石相结合。上海工匠的高超技艺令人惊叹，他们用水刷石做出很多漂亮的细部，仿石水平非常高，铁饰也很精致。东楼的轮廓线如此美丽，已经成为上海的城市地标。"唐玉恩评价道。

她认为这座造型优美的建筑作为文化建筑很合适，无论作为图书馆、美术馆还是历史博物馆，市民都很喜欢。不过，为上海历史博物馆找到这个合适的"新家"，却花费了原馆长张岚和团队数年的时间。

"我们做过好几个方案，包括汉口路原工部局大楼、外滩原汇丰银行、大世界、杨浦水厂和上海世博会城市足迹馆等。这些备选建筑由于面积、停车和安置费用等问题都不适合。"张岚回忆道。

上海历史博物馆源于20世纪30年代上海市博物馆的历史文献展厅和上海通志馆。1954年，上海市政府学习苏联模式筹建上海历史与建设博物馆，将老上海市博物馆的部分文物和上海通志馆的文献合并。长期以来，上海历史博物馆并无固定展陈空间，曾在文化广场、农展馆和东方明珠等地设展。

2010上海世博会举办后，上海美术馆迁往中华艺术宫，空置下来的昔日跑马总会大楼成为历博的"新家"。张岚认为2016年启动的改造工程非同一般，"是在一个保护建筑里建博物馆"。在满足功能的前提下，最重要的是保护好历史建筑。而工程的亮点把露出原来封闭的历史细部，如砖墙和藻井，展陈设计的线条也尽量与新古典主义线条吻合，

做到与老建筑完美结合。

唐玉恩提到，木雕、石雕和跑马主题特色的马头雕饰等细节都根据历史照片进行修复。工程进行中不断有新的历史细节被发现，设计方案也相应做了多次修改。

"经过修缮，东楼塔楼的大钟正常运行，给南京路沿线提供报时。我们对于历史建筑的保护和使用常怀敬畏之心，精心保护并科学使用，真正让历史建筑有尊严地走向未来。"她说。

虽然坐落于一座历史建筑里，上海历史博物馆却是按照现代博物馆的要求设计的，有恒温恒湿系统、流通的展线和自动扶梯。舒适照明和展柜的低散射玻璃都让参观变成愉快的享受。

如今，东楼和西楼由地下连廊连接，两楼之间的空间设计为

一个内院，保留着几棵历史悠久的高大树木。

"上海历史博物馆不是以专家作为参观主体，它为每一个对上海这座城市感兴趣的人而设计。观众们可以得到不同的享受，这是个让青年人受教育、让中年人休闲和让老年人怀旧的地方。"张岚说。

1933年的那个春天，伯克尔主席致辞时讲到，安放奠基石标志上海总会新纪元的开始。2018年的春天，这座建筑又开启了上海城市历史的新时代。

昨天： 上海跑马总会俱乐部　**今天：** 上海历史博物馆（上海革命历史博物馆）

地址： 南京西路 325 号　**建造时间：** 1934 年

建筑师： 马海洋行 Messrs. Spence, Robinson and Partners

参观指南： 周二到周日全天 9 点到 5 点开馆，建议欣赏建筑内的历史细节，如楼梯栏杆的马头雕饰。这里的展览也生动展示了上海这座城市独特的历史。

On a spring day 85 years ago, A.W. Burkill, chairman of Shanghai Race Club, laid on behalf of his wife the cornerstone of the new stands and clubhouse at the Bubbling Well racecourse.

A large gathering witnessed the ceremony, and lines on the stone read: "This foundation stone was laid on 21st May, 1933, by Mrs Katherine Burkill and commemorates the demolition of the original grandstand erected in 1863 and the erection of this building."

The stone can still be seen on the facade of the building which reopened in 2018 as Shanghai History Museum after years of preparation and restoration.

The new museum displayed 1,000 exhibits handpicked from its rich collection of some 110,000 antiques and documents on the history of Shanghai. However, Tang Yu'en, chief architect of the restoration, says "the largest exhibit is the building itself"。

"The building was built as Shanghai Race Club, which underwent changes and was used as

Shanghai Museum, Shanghai Library and Shanghai Art Museum. This is a monument that mirrors the history of Shanghai," Tang says.

When completed in 1934, the structure was described as "the finest and most luxurious in the Far East" by *the China Press* newspaper.

Tang says it is a typical club building, a unique genre of architecture in old Shanghai, which served as a social center for expatriates and showcased Western culture's spread to the Far East.

Several countries established their own clubs in Shanghai at that time, including Britain's Shanghai club and Germany's Club Concordia on the Bund, an American club on Fuzhou Road and two French clubs which are today's Science Hall and Okura Garden Hotel.

"It was rare for a Chinese city like Shanghai to have such a congregation of foreign club buildings in the downtown area. With a variety of inner spaces for multiple functions, these clubs served as high-end social centers, offering their members different services - from bars, restaurants, billiard rooms to reading rooms," Tang says.

The racecourse club was no exception according to *the China*

Press :

"On the ground floor are betting facilities for members and guests. On the first floor are recreation rooms, reading rooms, members' coffee room, guest rooms and card rooms. The upper floors contain two badminton courts, two English and four American-style bowling alleys, several squash courts and residential flats. The pavilion is made to accommodate a large number of spectators under one roof."

Tang says foreign club buildings, which differed from commercial architecture, were often designed in a more relaxed country style or in a style similar to the nation's traditional architecture.

Today, the race club is called the "east building" because the museum compound has a "west building", a former stable built in the 1920s which has been restored as an exhibition hall.

Both buildings have beautiful proportions and feature the British neoclassical style popular at the time. Unlike the sumptuous financial buildings on the Bund, the two race club buildings are in a more simple style. Instead of using a lot of stones, the architect combined red bricks with Shanghai plaster which was more economic.

"I was amazed at the superb skills of Shanghai craftsmen who created many beautiful details with plaster and imitated stone effects successfully. There are also delicate cast iron decorations. The east building has such a beautiful, lively shape whose silhouette has become a landmark of Shanghai," Tang adds.

She says the building is ideal for a cultural venue popular with Shanghai citizens no matter if it was as a library, an art museum or the soon-to-be history museum.

But it took years for former curator Zhang Lan and his team to secure the historic building which became vacant only after Shanghai Art Museum moved to the China Pavilion after World Expo., Shanghai 2010.

"Before this building was available, we had several plans and choices for the museum's new home - from the former Shanghai Municipal Council building on Hankou Road, the old Shanghai Natural History Museum on Yan'an Road to the former HSBC building on the Bund, Yangpu Water Plant and the Urban Footprint Pavilion of Shanghai Expo.,x s These plans all failed due to various reasons such as building size, parking places or relocation costs.

Finally the art museum became vacant," recalls Zhang.

The history of the race club building dates back to an exhibition hall of historical documents of Shanghai Museum in the 1930s and later a Shanghai history section organized by the city's cultural relics management committee. Without a permanent display space for decades, the museum had to host exhibitions in various places including the Oriental Pearl TV Tower in the Pudong New Area. The restoration project for the new home kicked off in 2016.

"It's a unique project because we built a museum inside a historical building, so preservation came first. It took great efforts to fit new functions of the museum into an old building," Zhang says.

The highlight of the restoration is the display of some hidden original parts, such as the exquisite caisson ceilings and brick walls. A galaxy of architectural details, from decorative sculptures on the verandas and wooden dados to ornaments shaped like horses' heads have all been carefully restored according to historical and field surveys.

"The restored bell on the tower top will work and give time to the Nanjing Road area, just like the bell tower of Custom House on the Bund. The principle of my work is to first respect old buildings which contain historical and artistic values, and then use them in a scientific way. So that they can be sustainably used and live in a new era with dignity," Tang says.

Though in a historical building, Shanghai History Museum is equipped with modern facilities such as temperature-and-humidity-control systems, escalators, comfortable lighting and anti-reflective glass to ensure a pleasant visit. Most of the exhibition content is child-friendly and interactive.

The east and west buildings are now linked by an underground corridor while the space between them will be turned into a garden with the original tall trees preserved.

"This museum is not only for experts and researchers, but for everyone interested in Shanghai. I hope it will be a place where young people can learn, the middle-aged can relax and more elderly visitors can enjoy some nostalgia," says Zhang.

In 1933, Burkill said the laying of the foundation stone marked a new epoch in the history of Shanghai Race Club. In the spring of 2018, the building marked the beginning of another new epoch

in the city's history.

Yesterday: Shanghai Race Club Building
Today: Shanghai History Museum/Shanghai Revolution Museum
Address: 325 Nanjing Rd W.   **Date of construction:** 1934
Architect: Messrs Spence, Robinson and Partners
Tips: The museum opens daily, except for Mondays, from 9am to 5pm (no admission after 4pm).

卡尔登公寓的半生缘

Eileen Chang's Last Home in Shanghai

长江公寓位于人民广场的黄金地段。这座建于上海公寓建筑黄金时代的大楼，也是著名作家张爱玲在上海的最后一个家。

长江公寓原名卡尔登公寓，是一座摩登的高层公寓。根据同济大学钱宗灏教授的研究，上海租界早期的住宅多为石库门——一种为小刀会起义后涌入租界的移民建造的联排建筑。根据中国人的习惯，石库门设有天井和厢房。20世纪20年代上海出现成套的里弄公寓，钢窗蜡地板与煤卫齐全，现代舒适。

"此后上海的地价不断上涨，开发商开始兴建6-9层的高层公寓。虽然高层公寓中每套平均房的面积更小，但因为更现代的设计和设施，居住品质反而提升了。"钱教授说。

卡尔登公寓就是一例。大楼建于1935年，浅褐色砖铺就的立面设计简洁，仅装饰有几个弧形的长阳台。卡尔登公寓和许多市中心的老上海公寓楼一样，底层为商铺，楼上是公寓。

根据1929年英文《大陆报》报道，卡尔登公寓属于名为"卡尔登地产"的综合开发项目。该项目位于今南京西路黄河路路口，除了卡尔登公寓，还包括卡尔登大戏院、美艺公司的产业和商铺等。卡尔登公寓原计划建10-12层，最终建了8层。

《上海老公寓》一书提到，上海的公寓楼多位于原公共租界和法租界的商业区。一些公寓楼后来成为地标建筑，如瑞金大楼、武康大楼和淮海大楼。

早期公寓建筑设计为古典风格和文艺复兴风格，20世纪20年代晚期后公寓建筑转向更现代的风格。到了20世纪30年代，高层公寓变得十分流行。1931年《大陆报》专门报道，"上海终于快速地成为一座公寓楼之城"。

钱教授提到，上海公寓建筑始于从20世纪20年代的建造活动一直持续到40年代。1949年后，上海政府兴建了许多工人新村，如曹杨二村。这些工人新村在老上海公寓的基础上加入苏联特点，配套建有学校、菜场和邮局。

在工人新村开始兴建的20世纪50年代，张爱玲从南京路另一

头的爱林登公寓（今常德公寓）搬到卡尔登公寓。当时的她已与胡兰成离婚。

研究张爱玲上海足迹的作家淳子发现，张爱玲很会挑地段和房子，她住过的房子都成为上海的优秀历史建筑。卡尔登公寓体量大，有四架电梯，楼梯铺有厚地毯，后面还有一个嵌入式的花园。公寓的地段无敌，附近有国际饭店、卡尔登剧院和大光明电影院，上海跑马厅和南京路商业街步行5分钟可达。

淳子认为，卡尔登公寓附近活色生香的市井生活也许给了张爱玲创作灵感。她和姑姑住在公寓301室期间，创作了电影剧本《不了情》《太太万岁》，小说《十八春》《小艾》。由《十八春》改写的《半生缘》被拍成多部深受欢迎的影视剧。她提到，张爱玲关于卡尔登公寓的回忆是由食物的味道引起的。

"在上海我们家隔壁就是战时天津新搬来的起士林咖啡馆，每天黎明制面包，拉起嗅觉的警报，一股喷香的浩然之气破空而来……只有他家有一种方角德国面包，外皮相当厚而脆，中心微湿，是普通面包中的极品，与美国加了防腐剂的软绵绵的枕头面包不可同日而语。"张爱玲写道。

1952年这位女作家离开了卡

尔登公寓,先赴香港后到美国定居。她再也没有回来,但上海成为她作品中反复出现的城市。

在张爱玲搬入卡尔登公寓前,这里居住着许多外国侨民。1941年3月5日,《北华捷报》刊登的一则新闻勾勒出公寓居民的生动肖像。

报道提到,加提亚先生(B. P. Ghatlia)的中国仆人被杀害,而他也被两个疑似日本人袭击。案件发生地点是卡尔登公寓。

"根据老闸巡捕房的调查,两个东方人在昨天早上7点半进入受害人位于卡尔登公寓125室的寓所。他们遇到受害人后,一人拿出匕首刺向加提亚先生,后者努力进行防卫,另一位闯入者袭击了他,后来这名印度人从2楼摔下,受了轻伤,引起众人围观。一些人了解情况后赶快上楼去他的房间查看,他们打开餐厅的门后看到不愉快的景象。房间中央躺着中国仆人,已经死亡了。他的头部几处都受到击打。"报道提到。这位可怜的仆人在主人受袭前已经死亡数小时。加提亚先生前晚外出回家很晚,没有看见仆人就去睡觉了。第二天一早,他被袭击他的闯入者惊醒。

报道并未提到作案动机,但透露加提亚在位于汉密尔顿大楼162室的一家贸易公司担任经理。这位印度商人后来到公济医院治疗,恢复不错。从自家窗口跳出逃生的他,只受了轻微的腿伤。

这篇惊心动魄的报道距今已有半个多世纪之遥,但报道中提到的汉密尔顿大楼、老闸巡捕房、公济医院和张爱玲享受面包香的卡尔登公寓,都幸运留存至今,成为上海这座城市宝贵的建筑遗产。

昨天: 卡尔登公寓 **今天:** 长江公寓 **地址:** 黄河路 65 号 **建造年代:** 1935 年
参观指南: 大楼隐藏了一个安静的院落。

Changjiang Apartments sits on a prominent location behind the Park Hotel at People's Square. Built during the golden era (1920s-1940s) of modern Shanghai apartment buildings, it's known as the last Shanghai home of famous author Eileen Chang.

"Changjiang Apartments, formerly called Carlton Apartments, was one of the many tall apartments which upgraded the city's residential life," says Tongji University professor Qian Zonghao, an expert in Shanghai's architectural history.

According to his research, the city's early residential buildings were mostly shikumen, or stonegate houses which were built after refugees from neighboring provinces flooded into the foreign settlement of Shanghai following the Small Sword Uprising in the 1850s. New-style three-story "lane apartments" emerged in the 1920s. Every flat in a lane apartment was en suite, equipped with a kitchen, a toilet, steel-framed windows and waxed wooden floors.

"As Shanghai's land price continued to soar since the 1920s, taller apartment buildings up to six to nine floors were built. Though every flat in taller build-

ings was smaller than before, living quality of flat dwellers improved due to modern design and facilities. Tall apartment buildings were in either art deco or modern style," Qian says.

Built in 1935, the Carlton Apartments was designed in such a simple-cut way that there are only several long, curved balconies gracing the facade. The ground floor features shops and the floors above are residential. The flank is painted in a warm chocolate tone, chic and modern.

According to a 1929 report in *the China Press*, the project for constructing Carlton Apartments was part of a large property transaction. Known as the "Carlton Property" situated at the corner of Bubbling Well Road (today's Nanjing Road W.) and Park Road (today's Huanghe Road) facing the former race course, the site included buildings such as the old Grand Theater, the Carlton Theater, the late Palais de Danse, premises occupied by Arts and Crafts, shops and houses. The initial plan for the now eight-floor Carlton Apartments was taller, up to 10 to 12 floors.

According to the book "Old Shanghai Classic Apartments," Shanghai's apartment buildings

were mostly located in commercial areas in the former international settlement or former French concession, which became signature buildings of the area such as Ruijin Building, Wukang Building and Huaihai Building.

Early apartment buildings were in classic or renaissance style, whose facades were embellished with architraves and carved brick decorations. Most apartment buildings changed toward the more modern style and simple-cut form since the late 1920s. In the 1930s taller apartment buildings such as the Carlton Apartments were prevailing in Shanghai.

A report in *the China Press* in 1931 also notes that "Shanghai has finally rapidly become a city of apartment houses."

The construction of apartment buildings in Shanghai started from the 1920s and continued until the 1940s. After 1949 the government built many residential compounds for local workers, the style of which were on the base of old Shanghai apartment buildings but was added with characteristic of similar residential compounds in the former Soviet Union.

After divorcing her husband

Hu Lancheng, Eileen Chang moved into the Carlton Apartments from Eddington House (today's Changde Apartments) near the western end of Nanjing Road in 1950. Chun Zi, a Shanghai author who has written a series of books on Eileen Chang's Shanghai footprints, finds Chang had favored apartment life and was good at choosing homes.

"The former Shanghai Race Course and the shopping streets — Nanjing Road and Fuzhou Road — were all within five minutes' walk away from the Carlton Apartments. I guess the busy urban scenes had inspired Chang during her city walks from the new home," she says.

In Flat 301 of Carlton Apartments, which Chang shared with her aunt, Chang completed several novels and movie scripts including *Eighteen Springs*, which she revised to a well-known novel *Affair of Half a Lifetime* in the US in the late 1970s and early 1980s. The novel was later adapted to a popular movie and TV series.

"Many years later, Chang's memories of life in the Carlton Apartments were about food aroma," she adds. "She wrote about the nice smells of bread fresh from oven of Kiessling Cafe in early mornings, which was moved from Tianjin during World War II," Chun Zi says.

Eileen Chang left the Carlton Apartments in 1952 for Hong Kong and later the US. She never returned to the city but Shanghai was forever a theme in her writings.

Before Chang moved in, the Carlton Apartments were home to many expatriate merchants. A news report of a murder in *the North-China Herald* on March 5, 1941 painted a picture of living in the Carlton Apartments.

According to the report, Wang Wei-chuan, Chinese servant of Mr B. P. Ghatlia, was hacked to death and he himself was attacked by two men believed to be Japanese, at the Carlton Apartments at the corner of Park and Burkill roads (today's Huanghe and Fengyang roads) on the morning of March 4.

The report said: "According to investigations conducted by detectives of the Sinza Police Station, the two Orientals entered the victim's flat No. 125 of the Carlton Apartments at about 7:30am yesterday and confronted him. One of the intruders pulled out a dagger and lunged at Mr Ghatlia who attempted to defend himself as best as he could. The other

assailant, however, also attacked the victim. The latter managed to fight them off and seeing no chance of escape but the open window, ran toward it and jumped out. The fall was from the first floor and a crowd gathered around the slightly injured Indian. Some persons, on learning of the occurrence, ran up and entered his flat. There they forced the door of the dining room open and saw an unpleasant sight. In the middle of the room lay the Chinese servant, apparently dead. His head had been hit in several places."

The police found the servant had died several hours before the attack on his master. Further investigations disclosed that Ghatlia was out the previous night and did not return until late. When he came home, he did not see his boy and went to bed. In the morning, he was awakened by the intruders who attacked him.

The report did not give motives or identities of the attackers, but said Ghatlia was the manager of M. Bhaichand, an import and export firm occupying room 162 in Hamilton House, which still sits on the corner of Fuzhou and Jiangxi roads. He was progressing favorably at the General Hospital. Jumping from the window of his flat, Ghatlia sustained slight injuries to his legs.

More than half a century has passed since that news report. Today, the Hamilton House where Ghatlia had worked, the Sinza Police Station which sent detectives to investigate the murder case, the General Hospital where Ghatlia were treated, and the Carlton Apartments where Eileen Chang enjoyed the aroma of the bread fresh from the oven of Kiessling Cafe — all survive to be heritage buildings and sites in Shanghai.

Yesterday: Carlton Apartments   **Today:** Changjiang Apartments
Address: 65 Huanghe Road   **Year of built:** 1935
Tips: The building conceals a quiet, nice yard inside.

上面这张照片摄于1920年代初。这是上海第一座公寓大楼，由Dollar Company在霞飞路（今淮海路）和亚尔培路（今陕西南路）转角处建造。

与上海相比，世界上只有两个城市的建筑发生了相同的变化，分别是洛杉矶和底特律。然而，在这两个城市中，都没有像黄浦江畔那样逆转了本地建筑风格。在1920年之前的半个世纪中，住宅的标准形式是蓝色和灰色的砖结构建筑。当时，一扇门通往很多家庭的住宅是很罕见的。到20世纪20年代，蒸汽供暖、现代化的抽水马桶和公寓房大约在同一时间出现。奇怪的是，英国人在自己的国家里完全依赖老式壁炉，在那里身体的前面烤得很热，后背仍然冰凉。而在上海，他们却是所有国家里最坚决要求供暖的。英国人还与美国人竞争，宁愿选择公寓式的居住方式，也不愿使用层高很高、房间多的房子。大多数公寓楼都是在上海建造的，这座城市的天际线已经完全改变。全世界任何地方都找不到比上海这里更舒适和时髦的公寓了。

The above photo, taken in the early 1920s, shows the first apartment building in Shanghai, built by officials of the Dollar Company at the corner of Avenue Joffre and Roi de Albert.

There are only two cities in the world where there has been an equal change in the building situated compared to Shanghai—namely Los Angeles and Detroit. In neither of those cities, however, has there been such a reversal of local architectural styles as there has been here on the banks of the Whangpoo. For half a century previous to 1920 the standardized form of dwelling house was a blue and grey brick structure. A door through which more than one family would enter their homes was indeed a rarity. Steam heat, modern flush toilets and apartment houses all came in about the same time, and curiously enough the British population, which in its own country is so thoroughly wedded to the old-fashioned fireplace, where you toast the front half of your body and freeze the rear part, has been the most insistent of all nationalities in Shanghai in demanding steam heat. The British have also vied with Americans in preferring the apartment style of living to the old-fashioned high-ceilinged house with many rooms, but mostly was built in Shanghai, the skyline has been entirely changed. Nowhere in the world are more comfortable and up-to-date apartment houses to be found than here in Shanghai.

摘自 1935 年 9 月 14 日《密勒氏评论》

Excerpt from *the China Weekly Review*, on September 14, 1935

哈沙德的菱形图案

Elliott Hazzard's Drape Patterns

1928年夏天，西侨青年会在上海跑马场对面开业。从那时起，这座大楼一直与体育运动紧密相连，成为"世界冠军的摇篮"。

西侨青年会是为旅沪外侨青年设计的住所，融娱乐设施和宿舍为一体，还有一座上海较早的温水游泳池。

1844年，基督教青年会由英国人乔治·威廉姆斯创办。1900年，中国基督教青年会在上海创办，1931年在不远处的西藏路兴建了一座融合中国元素的现代建筑。而建造这座西侨青年会大楼的想法源于1920年上海外侨精英的一次会议。

根据1928年3月24日英文《密勒氏评论报》报道，西侨青年会时任秘书长乔治·费奇提到与会人员认为要"给远离故乡的男士们一个家"，让"来自不同西方国家的男青年们互相认识，在健康的友谊中互相融入，在有吸引力的环境中舒适生活，而所需花费要在能够承受的范围内"。

这个想法在获得来自纽约和上海的捐赠支持后变为现实。由美国建筑师哈沙德设计的西侨青年会大楼高达10层，呈现意大利文艺复兴风格。大楼的现代化让媒体记者感到惊叹，认为是科学和金钱能做到的极致。

"南立面的粗糙墙面装饰有棕色、米色和奶油色砖拼成的菱形图案，这是一种让人印象深刻的宜人设计。底部三层的米色和棕色和谐融合，与土褐色的建筑形成好看的对比。"1928年6月30日的英文《大陆报》报道。

在这家英文报纸记者看来，大楼诸多建筑细节中，位于静安寺路（今南京西路）上的入口表现了这座建筑的重要性。入口有三座大门，门的上方设计有三扇高而缩进的窗，由两对修长优雅的柱式分隔。

美国密歇根大学研究员梁庄爱伦（Ellen Johnston Laing）发现，哈沙德为大楼立面设计了威尼斯主题的装饰。"底部数层用不同颜色的砖拼出菱形方块的图案覆盖整个立面，这一处理手法源于威尼斯总督府。把彩色砖呈几何图样地铺设在建筑表面，将

其从一个平淡无奇的墙面转化为令人激动的、精致的、蕾丝状的结构。"她在一篇关于哈沙德的论文中写道。两年后，哈沙德将这种装饰手法运用到上海新光大戏院的立面上。

哈沙德是老上海最重要的建筑师之一，1879年出生于南卡罗来纳州一个大米种植园家庭，在格鲁吉亚技术学院学习建筑。他毕业后在纽约知名的建筑事务所工作，1905年就在第五大道成立了自己的事务所。在纽约的建筑活动为哈沙德日后在上海的成就埋下伏笔。他1921年来到上海后，将当时风靡美国的布扎美术风格、意大利文艺复兴风格和英式风格等移植到了上海的建筑风貌中。

1923年上海英文报纸的广告里提到，上海只有三位"真正的美国建筑师"，他们分别是哈沙德、东欧建筑师邬达克早期的合伙人克利（R.A. Curry）和后来设计北大、清华校园的茂飞(Henry Murphy)。

有趣的是哈沙德初到上海时曾与茂飞共事，还居住在茂飞位

于外滩3号顶楼的公寓里。1923年茂飞关闭上海办事处后，哈沙德留在上海独立执业。他还设计了很多知名作品，如金门饭店（华安大楼）、枕流公寓、静安宾馆和淮海西路富兰克林住宅。

哈沙德设计的西侨青年会和华安大楼落成后，成为现代上海摩天楼的标志性建筑。两座大楼雄踞在跑马场的一侧，控制着上海跑马厅一带的天际线。不过南京路的地段升值太快，两座建筑的霸主地位并未保持很久，1934年邬达克设计的国际饭店作为中国最高的建筑成为新的视觉中心。而今天再看，国际饭店的霸主位置又被超越，棕褐色颀长的低调身影隐没在摩天巨厦的海洋里。

在近年网红的"上生·新所"城市更新项目中，哈沙德设计的哥伦比亚俱乐部又和邬达克作品孙科故居比邻，被一些媒体混淆为邬达克作品。

对比哈沙德和邬达克的人生轨迹，也有一点"既生瑜，何生亮"的味道。他们年龄相差十几岁，在3年间先后到沪发展。从职业资历看，哈沙德在纽约等大都市有不少建筑实践，比大学毕业就入伍的邬达克更胜一筹。两人到上海后都赶上了这座城市的黄金建设时代，留下很多经典作品。抗战后他们都选择留在上海，事务所一度门可罗雀。1943年哈沙德在日军集中营病逝，而邬达克1947年离开上海后去美国定居。两人都活了60多岁。

当年的英文报纸提到，西侨青年会大楼的室内设计与外立面一样，美观实用。大楼中央供暖，装修精美的房间光线充足、空气新鲜，健身房、游泳池、跑道、保龄球道、俱乐部和更衣室一应俱全。西侨青年会还提供优质餐饮、学习课程、研究设施和拓展讲座。所有房间的龙头都提供过滤水。炎炎夏日里，水会通过冷却装置冰镇，而健身房角落里也有汽水桶随时提供饮料。

抗战时期，大楼被日军占

领，1950年由上海市政府接管。在时任上海市长陈毅的建议下，西侨青年会大楼改为上海体育俱乐部，1957年开始对儿童和青少年开展游泳、围棋和国际象棋的业余训练，从中发现培养体育人才。从那时起，西侨青年会变成一座"冠军的摇篮"，许多上海籍的世界冠军，如游泳选手杨文意和围棋选手常昊都是从体育俱乐部开始启蒙训练的。

如今，已改名为"体育大厦"的大楼开设了上海体育博物馆，展示中国参与奥运会的历史。一块块奥运金牌和成长于"体育摇篮"的冠军照片，与哈沙德设计的华丽天花板相映生辉。

上海体育俱乐部主任梁立刚认为，这座楼保护得比较好，因为1949年后一直是体育局使用，没有更换过单位。"从局领导到每个员工，都对这座设计精美的历史建筑很有感情。有些老员工回忆，大楼夏天可以不用空调，因为对流特别好，到处回响着风吹关门的声音。"他说。

体育大厦的很多室内细节，也保持着90多年前开幕时的旧貌——如入口处的三座门、深木装饰的图书馆、温水游泳池，还有哈沙德的菱形图案。

昨天：西侨青年会大楼　**今天：**体育大厦　**建造时间：**1928 年
建筑师：哈沙德　**建筑风格：**意大利文艺复兴风格
参观指南：建议参观位于二楼的上海体育博物馆，每周二到周日上午 9:30 到 11 点，下午 2 点到 5 点对外开放。体育大厦的门厅和二楼博物馆大厅都有大量哈沙德设计的建筑细节。

The former Foreign YMCA building opened in the summer of 1928 opposite Shanghai Race Course, today's People Square. Since then the building has always been linked with sporting activities and later became "a cradle of world champions".

The building was constructed to provide accommodation and recreation for foreign young men in Shanghai. The idea of building this magnificent building originated in 1920 when a group of leading men in the city discussed the formation of a Foreign Young Men's Christian Association in Shanghai.

The history of the YMCA dates back to 1844 when Englishman George Williams founded the international organization. In 1900 a Chinese YMCA was found in Shanghai and a modern building with Chinese elements was also built on the People's Square.

In an article in *the China Weekly Review* on March 24, 1928, the then Foreign YMCA general secretary George Fitch recalled it was to provide "a home for men away from home" and "a place where young men of various Western nations could meet and mingle in wholesome fellowship and where they could live comfortably in at-tractive surroundings at a cost of what would be within the reach of all".

With generous contributions from both New York and Shanghai, the idea turned into reality — a fire-proof building designed by American architect Elliott Hazzard.

Heading an influential architectural office in Shanghai in the late 1920s and early 1930s, Hazzard designed three other buildings along Nanjing Road, including the Wing On Tower, Shanghai Power Company and the adjacent China United Assurance Building. The latter and the Foreign YMCA building had dominated the skyline of the former Race Course before the erection of the Park Hotel in 1934.

The 10-floor Foreign YMCA Building in Italian Renaissance style was described by local media as a wonder building "as modern as science and money can make it".

"The rusticated walls of the lower story on the southern facade were in a diaper pattern of brown and rich buff and cream brick set to an effective and pleasing design. These three lower stories blend well with the buffs and the browns and form a pleasing

contrast to buildings of drab coloring. The walls have been broken up by bringing out piers between windows and at the corners which carried right up to the red-tile caps. The piers at the corners are buttressed slightly and the effect secured adds greatly to the charm of the building," the China Press reported on June 30, 1928.

In the eyes of this reporter from the American newspaper, it was the triple entrance on Bubbling Well Road (today's Nanjing Road W.) with its three tall and deeply recessed windows above separated by two pairs of slender and graceful columns that gave significance to the architecture of the building among the many details.

Ellen Johnston Laing, a researcher from Michigan State University, noted the drape pattern was a treatment formerly used on the famous Palazzo Ducale di Venezia. In a study about architect Elliott Hazzard, she wrote that this decorative, very special treatment on the facade was rarely seen in Shanghai. Hazzard used it again in the Xinguang Theater on Ningbo Road two years later.

And the interior of the building was in keeping with its exterior which was not only attractive but

also rather functional. Although an institution with a religious background, both the Chinese YMCA and the Foreign YMCA carried out various activities focusing on young people including sports activities and informative lectures.

The building provided centrally heated, fully furnished rooms with plenty of light and air, a gymnasium, a swimming pool, running track, bowling alley, good meals, club rooms, dressing rooms, special courses of study, facilities for research work, extension lectures and the like.

In all rooms, filtered water from the faucet was available. During the heat of summer the

water was first sent through a large cooling coil. Fountains in the corners of the gymnasium provided refreshment for athletes.

During World War II, the building was occupied by the Japanese army and in 1950 it was taken over by the Shanghai government. The then mayor, Chen Yi, assigned the building to Shanghai Sports Bureau, which opened the Shanghai Sports Club here in 1957 to organize start-up sports classes to select and train future athletes from local children and youngsters.

Since then it became "a cradle of world champions". A galaxy of Chinese stars, including swimmer Yang Wenyi and the go player Chang Hao, started their careers as amateurs at the Shanghai Sports Club.

Now part of the second floor is open to the public as an exhibition room of Shanghai Sports Museum. The history of the Olympic Games and Chinese participation is showcased. Olympic gold medals and huge pictures of smiling Chinese champions who had their first training here glisten under the gorgeous ceiling of this well-designed building.

"The building was well preserved because it has always been used by Shanghai Sports Bureau since 1950. From the bureau leaders to every staff member, we have deep feelings for this well-designed historical building. Some old staff recalled the ventilation was so good that air conditioners were not necessary in hot days. There was constantly the sound of wind blowing doors to close," says Liang Ligang, club director.

It's true that much of the original exterior and interior has remained as on the opening day 90 years ago, such as the triple entrance, the beautiful arches, the dark-wood library, the swimming pool still with warm water in winter and Elliott Harzzard's signature diaper patterns on the facade.

Yesterday: Foreign YMCA Building   **Today:** Shanghai Sports Club   **Built in** 1928.
Architect: Elliott Hazzard   **Architectural style:** Italian Renaissance style
Tips: The exhibition room of Shanghai Sports Museum on the second floor is open to the public at 9:30am to 11am, 2pm to 5pm from Tuesdays to Sundays. The exhibition hall as well as the lobby feature abundant original architectural details.

赖斯上校在西侨青年会开幕式上的讲话
Colonel G. R. Rice's speech at the opening ceremony

我想提醒您，三角形是三边形的，代表了人格的三位一体——精神、灵魂和身体。建造这座建筑的目的明确，就是要照顾上海年轻人的这三重需求。这里有一个可供举办演讲和类似活动的高级大厅，为满足年轻人的精神和智力需求做好准备。其实不仅是年轻人，一些年长的人也强烈希望能受邀参加这里的活动。对于年轻人和年长者来说，这座建筑的演讲厅都可以成为传播一切美好事物的中心。那么，大楼为社交活动提供了哪些优质的设施呢？我敢肯定，在健康的氛围中，没有什么地方更适合为那些精神上和智力上有全面需求的人提供服务了。

我还需要讲"三角形"的那第三条边吗？身体的舒适感在这里并没有被忽略，体育馆、游泳池、餐厅、休息室、起居室等设施，可以满足身体的需求。

沃洛普将军希望引起注意的一点，是基督教青年会工作的国际性。"尽管这座特殊的建筑是专门为外国人建造的，但基督教青年会在各个种族间开展工作，强调全人类的兄弟情谊。在中国的年轻人中也做了大量的工作，而中国基督教青年会在为那些真正善良的中国人提供了一个聚集的地方。这座大楼也将为不同种族和国家的外国年轻人提供相遇的机会，可以讨论问题。在社交时间，不同国家的人可以一起喝杯茶，以友好的方式交流，解决他们脑海中所困扰的问题。并且随着不同观点的出现，人们能够通过他人的眼睛和视角来看待事物。为了人类共同的利益，他们将会取得进步。而彼此了解得越多，我们越有能力看到问题的两个方面。通过基督教青年会和此类机构建立的国际友谊具有真正的价值，我们有责任尽力协助执行"彼此相爱"的宗旨。

May I remind you, that the triangle is three-sided and it represents the trinity of the human personality — spirit, soul and body. This building has been put up with the express purpose of ministering to the three-fold needs of the young men of Shanghai. There is the fine hall which is available for lectures and the like, and it is here that provision is made for meeting the spiritual and intellectual needs of Shanghai's youth; and not only the youth, for it is the fervent hope of some of the older ones among us that we shall be invited to attend such addresses, and that the lecture hall of this building may be a center for the dissemination of all that is best for young and old...

Then what fine facilities are afforded for social intercourse! There is this side for which provision is required as well as the spiritual and intellectual side and I am sure that no place is better fitted for ministering in a healthy atmosphere to those necessities of a full-orbed existence than the building in which we now find ourselves.

Need I refer to the third side of the triangle? The comfort of the body has not been overlooked. Inspect the gymnasium, the swimming pool, the dining hall, the lounge, the living apartments, and see how admirably the needs of the body have been catered for.

There is one more point to which Gen. Wardrop wished to draw attention and that is the international character of the work of the YMCA. Although this particular building is specially for foreigners, I suppose we are aware that the YMCA carries on its work among all races and does all in its power to emphasize the brotherhood of man. There is a large work among the young men of China and the Chinese YMCA has done much to form a rallying point for those who have the real good of China at heart.

In this building, too, there will be given a common ground for foreign young men, not all of the same race, to meet and discuss questions which affect them. It will be possible during the social hour, over a cup of tea, for men of different nationalities to ventilate, in a friendly way, the problems that occupy their minds, and, as the different points of view are brought to light, and a man is enabled to see matters through the other man's spectacles, definite progress will be made in the interest of humanity, for the more we know one another, the more we understand one another, the better able are we to see that there are two sides to every question and that all the right is not on our own side. International friendships, brought about through the agency of the YMCA and such institutions, have a real value, and it is our duty to do all in our power to assist in the carrying out of the command of the Mater—"Love one another".

摘自 1928 年 7 月 7 日《北华捷报》
Excerpt from *The North-China Herald,* on July 7, 1928

一次中西合作的典范

A successful Sino-foreign cooperation

上海中西合璧的建筑故事很多，金门饭店是一个经典案例。1926年，一家华资保险公司投资，聘请美国建筑师哈沙德设计兴建了这座意大利文艺复兴风格的建筑。

上海档案馆研究员张姚俊认为，这是上海历史建筑中一座罕见的"保险大楼"。当时，保险公司多为银行所办，办公设在银行大楼中，很少兴建一座保险公司大楼。友邦保险的创始人史带是租用外滩字林西报大楼开展业务的。更值得一提的是，建造大楼的华安保险是中国第一家纯华资的保险公司，由保险业大佬吕岳泉创办。

张姚俊从上海档案馆馆藏的吕氏后人回忆中勾勒出吕岳泉传奇的发家史。吕岳泉是上海川沙人，儿时家贫，到英商永年人寿公司外籍经理穆勒家当帮佣。他聪明好学，对主人的保险经纪业务耳濡目染，还自学了外语。穆勒偶然发现吕岳泉的才能后请他当自己的助手，做翻译等工作，又推荐他到公司从事保险代理工作。

因为业绩好，吕岳泉升任了永年人寿南京分公司总经理。后来，他听从两江总督端方等人建议，创业开办了第一家纯华资的保险公司——华安合群。公司还引进加拿大精算师郁次等一批外籍人才，于1912年7月1日在外滩规矩会堂开业，后来又搬到南京路。

华安公司将一半的资金用于投资上海的地产项目，盈利可观。根据1923年美商《大陆报》报道，因为外籍高管郁次的建议，公司为新大楼在原跑马场对面选址购地。"很多人批评在那么远的地方做生意简直愚蠢，而在静安寺路（今南京西路）建那么宏伟的一座大楼真是疯狂的主意。"报道提到。不过此后，南京路地价飞涨，事实证明新大楼的选址是一笔非常明智的投资。

华安保险的钢筋混凝土大楼被誉为"上海最上等的建筑"，是一座不计成本的美厦。

大楼设计师——美国建筑师哈沙德创办的哈沙德洋行是当时颇具影响力的建筑事务所。哈沙德还设计了南京路上其他三座

标志性建筑——比邻的西侨青年会、更摩登现代的永安新厦和上海电力公司。在1934年邬达克的成名作国际饭店落成以前,哈沙德设计的华安大楼和西侨青年会曾经控制了上海跑马场(今人民广场)的天际线。

这座9层保险大楼高达76米,顶部有一座钟楼,底部两层以花岗岩贴面,其余楼层的混凝土外立面有石材效果。有研究认为华安大楼的灵感来源于费城独立纪念堂,而大楼落成的1926年是中华民国成立15周年,也是中国从封建统治下获得新生15年纪念。

大楼建成后,底部两层是商店、图书室和保险公司,其余楼层是公寓。《大陆报》报道评论,"这也许是世界上第一家由人寿保险公司经营的精装修公寓楼。"

1928年的报道还提到,"二楼的办公室用意大利大理石和硬木装饰,会议室和经理室饰有柚木。公寓都非常美丽,装饰奢华,且层高宜人,夏季可以享受不断地吹拂纯净清凉的微风。大

楼和餐厅由史达特曼夫人管理。她会在早上6点半前亲自去市场采购食物，并监督烹饪，以确保完美的餐食和服务。"

大楼共有68套公寓，145个房间。餐厅不仅为110名公寓租客提供服务，也向附近的客人开放。华安大楼的公寓深受欢迎，非常抢手，有着长长的等候名单。这是因为"在远东地区，很难找到这样拥有高品质食物和便利地段的住房"。

华安保险在鼎盛时期分支机构遍布南洋，1937年抗战爆发后业务开始走下坡路，后来不得不将大堂和大部分楼层租给金门饭店，仅留二层办公。金门饭店是近百位中国企业家合资于1936年在香港注册创办的，1937年开办上海分公司，1939年租赁华安保险大楼，次年12月30日开业。

1949年后，大楼由中国纺织局租用。1953年吕岳泉在香港病逝后，大楼由上海市政府接管，改为接待归国华侨的华侨饭店。1992年，酒店为吸引海外旅客恢复了金门饭店旧名。

如今，金门饭店有不少客人是喜欢酒店历史气息的外国游客。2017年上海市开展"阅读建筑"活动后，金门饭店把大堂的商务中心改为一间酒店历史陈列室。

如今，酒店大堂仍矗立着巨大的大理石柱，雕刻繁复精致。陈列室位于一个安静的角落，虽然面积不大，但历史照片述说了一间华资保险公司曾经的辉煌与骄傲，值得一看。

昨天：华安联合保险公司大楼　**今天：**金门饭店　**地址：**南京西路 108 号　**设计师：**哈沙德
参观指南：建议参观位于大堂的历史陈列室。

The China United Assurance building is living proof of a once successful Sino-foreign cooperation. Designed by American architect Elliott Hazzard, this stately Italian renaissance building was built in 1926 for a Chinese insurance company run on foreign lines.

"Among Shanghai's galaxy of historical buildings, this is a rare example of a building devoted entirely to insurance. At that time, most insurance companies were operated by banks and located inside a banking house. Even Cornelius Vander Starr, founder of American Asiatic Underwriters and forerunner of AIG, rented an office in *the North China Daily News* building on the Bund instead of building a headquarter," says Zhang Yaojun, a researcher from Shanghai Archives Bureau.

Owner of this edifice, China United Assurance Co. Ltd. was the first full Chinese-capital life insurance company founded by local man Lu Yuequan.

According to Zhang's research, Lu had very humble beginnings, born into to a poor family in Chuansha, Pudong, and working for a foreign manager of China Mutual, a British life insurance company.

Amazed at his young servant's self-taught English and his knowledge of insurance, the manager made Lu his assistant and later recommended him for a post at China Mutual.

"Owing to his remarkable performance, Lu was appointed as general manager to open the company's Nanjing branch. In Nanjing, capital of Jiangsu Province, he made friends with powerful political figures who persuaded him to found China United Assurance Co. Ltd. in 1912, which was the first Chinese life insurance company organized with purely Chinese capital and run by Chinese," Zhang says.

Because of his experience working in a foreign insurance company, Lu intended to run the company on foreign lines. He invited a few foreigners, including his former Canadian colleague, A.J. Hughes, to be the general manager and train Chinese staff. It was an experiment, but proved a suc-

cess.

The company opened on July 1, 1912, at the Masonic Hall on the Bund, later moving to Nanjing Road. The company had been fortunate in their investments from the start, and around half their funds were in Shanghai real estate.

According to a report in *the China Press*, in 1923 the company purchased the present site opposite the former racecourse at Hughes' suggestion, "although many critics thought it almost folly to locate a business so far out, and especially a wild idea to put up so great and fine a building on Bubbling Well Road".

It was a smart investment as land prices along today's Nanjing Road were soaring in the 1920s and early 1930s.

The building was a steel frame and reinforced concrete structure which was reviewed as "the choicest specimen of architecture in Shanghai" and "ranked with the first half dozen in attractive lines irrespective of cost".

Architect Hazzard, who headed an influential architectural office in Shanghai in the late 1920s and early 1930s, designed three other buildings along Nanjing Road, including the Foreign YMCA building and the more modern Wing On Tower and Shanghai Power Company.

The China United Assurance building and the adjacent Foreign YMCA building had dominated the skyline of the former racecourse before the erection of the Park Hotel in 1934.

The nine-floor insurance building was surmounted by a clock tower crowned with a lantern at a height of 250 feet (76 meters). The first two floors are faced with granite with imposing effect. The other floors, with the appearance of plain stone, are in reality concrete.

The offices of Lu's company occupied the first floor and there were two large vaults for books on the ground floor. The remainder of the ground floor was occupied by stores and shops, and all other floors had been furnished and served as apartments. The China Press called it "probably the only life insurance company in the world running a furnished apartment building".

Its 1928 report said: "The offices on the first floor are finished in Italian marble and hardwoods, the boardroom and the manager's room in teak. The apartments are also beautifully finished and luxuriously furnished, and the height

above ground gives a constant pure, cool breeze in summer. The management of the building itself is in the hands of Mrs M. Stadtmann, who also runs the dining room, making all purchases herself at the market before 6:30am. She superintends the cooking and in general sees to it that perfect meals are served and perfect service given."

There were up to 68 apartments and a total of 145 rooms in the building. The restaurant served not only the 110 tenants of the apartments, but also guests from nearby. There was always a waiting list for the rooms and apartments in the building as "it was not easy to duplicate the quality of the food, convenience and fine appointments of the rooms anywhere in the Far East".

The heyday of China United Assurance came to an end when the Chinese War against Japanese Aggression broke out in 1937. Their business all over the country almost came to a halt and branches were closing down. The company had to rent out the lobby and other floors to the Pacific Hotel, maintaining only the second floor as its office.

The Pacific Hotel Co. Ltd. was launched and invested by more than 100 Chinese entrepreneurs. The company was registered in Hong Kong in 1936, opened a Shanghai branch in 1937, rented the China United Assurance building in 1939 and opened the Pacific Hotel on December 30, 1940.

After 1949, the building was lent to the East China Textile Administration as its office. It was taken over by the Shanghai government after Lu died in Hong Kong in 1953 and since then reopened as the Huaqiao Hotel to host Chinese who returned from overseas. In 1992, the hotel reverted to its old name, the Pacific Hotel, to attract more overseas tourists.

Now around 40 percent of guests are foreigners who like the historical ambience of the hotel which now belongs to the Shanghai Jin Jiang International Hotels (Group) Ltd. In 2017 the hotel transformed the former business center on the ground floor into an exhibition room of history.

The exhibition room sits in a quiet corner of the hotel's lobby graced by tall marble columns and exquisite sculptures. The room is not big, but precious old photos and silver tableware showcase the history of a fine building built by a Chinese insurance company that is well worth a read.

Yesterday: China United Assurance Building   **Today:** The Pacific Hotel
Address: 108 Nanjing Rd W.   **Architect:** Elliott Hazzard
Architectural style: Italian Renaissance style
Tips: Please visit the history exhibition room with a bilingual introduction to the hotel's history.

华安大楼开幕仪式

Colonel G. R. Rice's speech at the opening ceremony

周日，华安联合保险公司位于静安寺路34号的新总部大楼在举行正式开幕仪式。大楼为此特别布置装饰，众多来宾出席了仪式。

当天主持活动的是徐固卿将军，他用中文概述了华安公司的历史。他说，1925年底，公司的有效业务达1,170万两白银，资产有176万两白银，去年偿付了176万两理赔，到期款项有43.9万两。

华安公司总经理兼名誉董事郁次（A. J. Hughes）在介绍上海工部局总董费信惇（Stirling Fessenden）时说："在实现华安保险公司宗旨的历程中，这座大楼的竣工是又一个里程碑。华安保险成立于1912年，就在中华民国成立后不久，而今天是中华民国成立15周年纪念。我们期望在新的秩序下做一些真正的具有建设性价值的事情。

很自然地，我们都为这座大楼感到自豪。我谨代表公司全体董事和管理层，借此机会郑重地表达我们对建筑师哈沙德先生的欣赏。他的作品本身就展示了他的专业技能。我们特别请哈沙德来设计这个项目，是因为他曾在纽约设计类似的建筑，有丰富而宝贵的经验。在整个设计和建造过程中，这种经验的好处非常明显。

我还想借此机会就大楼所在土地的产权情况辟谣。它不是租赁产权，而是永久产权，我们永久拥有这块地。这块地产包括一块近12亩地的方形方块，其中大楼和后方车库占地约3亩。这块土地于1923年11月以远低于现在市场价的价格购入。当时我们预期商业区将向西快速扩张，投资收益会高于中央区，目前这些预期已经实现。

The chairman of the day was Gen. Hsu Ko-ching who outlined in Chinese the history of the company, and said that, at the end of 1925, the society had Tls. 11,700,000 business in force and Tls. 1,760,000 in assets. It paid Tls. 1,713,000 in claims and Tls. 439,000 on maturity last year.

Mr A. J. Hughes, the general manager and honorary director, in introducing Mr Stirling Fessenden, Chairman of the Shanghai Municipal Council, said: "The completion of this building is another mile-stone on the road to the fulfillment of the aims of the promoters of the China United Assurance Society. The Society was started in 1912, just after the inauguration of the Chinese Republic, of which today is the 15th anniversary, in the hope of doing something of real constructive value for the new order of things.

Naturally we are all very proud of this building and on behalf of the Directors and Executive staff I wish to take this opportunity of expressing in a more public manner than has hitherto been possible our appreciation of our architect, Mr Elliott Hazzard, whose finished work speaks for itself of his professional skill. He was specially selected as having had a wide and valuable experience in the erection of similar buildings in New York and the benefit of that experience has been very much in evidence during the whole course of designing and construction.

I take this opportunity of disposing of a rumor which has reached us in regard to the land on which this building is erected. It is not leasehold but freehold property, our own in perpetuity. The whole consists of a square block of nearly 12 mow of which this building and the garages in rear occupy about three mow. It was purchased in November, 1923, at a figure very considerably less than the present market value in anticipation of a very rapid expansion of the commercial district westward and a much more profitable investment than the Central District afforded. Both these anticipations have been realized.

摘自 1926 年 10 月 16 日《北华捷报》
Excerpt from *the North-China Herald*, on October 16, 1926

深藏校园的"巡捕房"

位于南京路黄金地段的上海商贸旅游学校内，静默地矗立着一座19世纪的巡捕房。这座红砖大楼竣工于1889年，曾是老上海公共租界15个捕房之一的老闸巡捕房。大楼跨越一个多世纪，见证了公共租界警察制度的发展历史。

1845年英租界成立时并没有警察，只有负责火灾警报的华人更夫。1854年工部局成立，负责公共租界的市政管理。工部局很快就把租界警察局——巡捕房设立起来，以维持社会秩序和公共治安。

根据张彬所著的《上海英租界巡捕房制度及其运作研究》，上海英租界巡捕房是仿照19世纪20年代末出现的英格兰现当代警察制度建立的。

"上海英租界开辟之后的将近九年时间内，一直实行华洋分局政策，租界内人少事简。上海小刀会起义爆发后，上海地方政府自顾不暇，百姓则大量涌入英租界中，使租界局面为之一变。租界内的外侨为了自己的商业利益，不愿撤离上海。在战争环境下，他们同时又希望能保证租界的安全和秩序，于是决定成立租界自治政府。巡捕房也随之应运而生。"张彬写道。

为了尽快成立巡捕房，工部局从香港请来巡捕头，其雇佣的第一批巡捕普遍受教育程度低，缺乏专业度。19世纪末，工部局招募英国警校毕业生充实巡捕队伍，并陆续完善巡捕的养老退休制度后，情况大为改善。

工部局的15个捕房中，老闸巡捕房覆盖一块由苏州河、延安路、西藏路和陕西路围合的中心城区。1888年8月1日，耗资46000两白银的老闸巡捕房开工建设，于次年落成。其中，巡捕房的主楼和部分牢房保存至今，都深藏在商旅职校的校园里。

1889年12月6日，巡捕房竣工在即，英文《北华捷报》报道称"近期到访上海的人们为一座藏在难看的抛球场后面的红砖大楼那漂亮的立面感到惊叹。"报纸还解释，人们惊讶于如此壮观的大楼居然只是老闸区的巡捕房。

"主楼呈现威尼斯哥特风格，由红砖建成，饰以灰砖和宁波石。与很多背面和侧面使用普通砖或灰泥等廉价材料的红砖大楼不同，这座新警局建筑的每一面都是红砖砌成的。"报道写道。

大楼的一些墙面还用带质感的灰色涂料粉刷，视觉效果很好。门和栏杆扶手用柚木制成，坚实牢固。门口木饰和栏杆支柱等部位的装饰由中国工匠雕刻而成。楼梯的特色是位于三楼的拱廊，砖柱上方有三个拱券，柱头雕饰精美。

如今，大楼保留着很多19世纪报纸描述的原貌。占地936平方米的大楼气势威严，高达三层，每一层的前部都设计有美丽的外廊。

这座建筑不但美观，作为一

座容纳不同国籍巡捕的19世纪的警局，也相当实用。

《北华捷报》的报道提到，大楼一层主要是西捕房间、备餐室、餐厅、弹子房和华捕房间等。二楼分成两部分，建筑的一翼和部分中央空间由总捕头专用，其他部分由印度锡克族巡捕使用。总捕头拥有卧室、起居室、餐厅、浴室和备餐室各一间。在锡克族巡捕使用的一侧，有警官的卧室、一间用于点香的房间、翻译房间和一间大卧室。

三楼则由来自欧洲的警官和巡捕使用，中间是由12名巡捕共用的卧室和盥洗室。一侧是四名警官居住的一间带浴室卧室，另一侧则是带有浴室的阅读室。三楼通往顶部塔楼，在塔楼上俯瞰租界城市风貌视野绝佳。瞭望员还通过一根通话管道与底下的人沟通。

从今天的眼光来看，老上海国际化的巡捕队伍分工合作得很好。巡捕房由多为欧洲人的西捕领导管理，擅长包打听的华捕采集消息，印度锡克族的巡捕指挥交通，日籍巡捕则管理日本工厂

较多的虹口和普陀一带。1889年竣工的老闸巡捕房可以关押30名华犯、10名西犯和20名乞丐。

作为上海中心城区一个重要的警局，红砖大楼见证了很多近代历史事件，其中影响最大的是"五卅惨案"。1925年5月30日，租界当局拘捕了多名爱国学生，关押在老闸巡捕房。愤怒的群众聚集在捕房门口，要求释放被捕学生。英国巡捕开枪射击，造成9人死亡，14人受伤，成为震惊中外的"五卅惨案"。如今，"五卅"纪念碑就在人民广场，离惨案发生地的红砖大楼步行只有几分钟之遥。

1949年后，昔日巡捕房由培光中学使用，经过多次并校，1985年改为上海商业职业技术学校，2006年更名为上海市商贸旅游学校，以培养旅游专业人才为特色。

李小华校长回忆，老闸捕房曾被白色瓷砖包裹多年，2009年大修时工人敲开瓷砖，露出红砖立面时大家感到"很震撼"。学校为此增加了经费，保留并修复了红砖立面、木窗和壁炉。遗

憾的是，视野绝佳的塔楼因改建需要被拆除了。学校还保留了约100平方米的老监狱。

如今，这座19世纪的巡捕房成为学校的标志性建筑。底楼开设了一个迷你博物馆，展示大楼的建筑历史和上海商业史。楼上用作会议和学生社团活动。红砖大楼对面是昔日牢房，边上是学校的"咖啡文化驿站"，学生们在此学习制作各色饮料。

李校长提到，学校还有文博专业。"我们将在这座历史建筑里举办音乐会，也计划让文博专业的学生做关于大楼历史的调研，并设计制作建筑铭牌。用这种方式，历史文化遗产的价值可以传承给年轻一代。"他说。

昨天：老闸巡捕房　**今天：**上海市商贸旅游学校　**地址：**贵州路 101 号

建筑师：Arthur Dallas　**建筑风格：**威尼斯哥特风格

参观指南：建筑不对外开放，但从校门口可以欣赏这座 19 世纪建筑精美的红砖立面。

A red-brick building on the campus of Shanghai Business & Tourism School is a witness to the history of policing in old Shanghai. Completed in 1889, the Louza Police Station on Nanjing Road was the first of the 15 branches of the then Shanghai Municipal Council's Central Police Station.

When founded in 1845, the former British settlement had no police officers but only Chinese nightwatchmen to look out for fires. After the Shanghai Municipal Council that managed the foreign settlements was established in 1854, the police station was formed immediately to maintain social order and public security.

According to Zhang Bin's book *A Study of the Police System and Its Management in the Former Shanghai British Settlement (1854-1863)*, Shanghai's police organization directly referred to the English police system which originated in the 1820s.

"During the first decade of the former British settlement, foreigners and Chinese lived separately. There were only a few foreign residents and civil affairs. After the Small Sword Society Uprising in 1853, some 20,000 Chinese refugees flooded into the settlement. The expatriates, who profited from leasing houses to the Chinese, found emerging problems when refugees drank a lot, fought and caused traffic jams. So there was an urgent need to restore social order," Zhang wrote in the book.

That led to the immediate founding of the Shanghai Municipal Council and its police station in 1854. At the beginning, the constables and policemen were neither well-educated nor professional. But when it came to the late 19th century, the local police team improved greatly after the council recruited graduates from British police schools and sent constables for training in Ireland.

Louza Police Station was established to cover a central area enclosed by today's Suzhou Creek and Yan'an, Xizang and Shaanxi roads. The project, which cost 46,000 taels of silver, kicked off on August 1, 1888 and was completed the following year. The main building, a brick-and-concrete structure, as well as a few old cells are well preserved today.

When the police station was nearing completion on December 6, 1889, *the North China Herald* reported that "recent visitors to Shanghai have noted with some astonishment the handsome facade of a fine red-brick building nearly hidden behind the unbeautiful Racquet Court." The paper explained that what surprised visitors was the fact that such an imposing building, "the first really important public edifice constructed by the municipal council," was simply a police station for the Louza district.

"The main building is mainly in the Venetian Gothic style and is of red brick with dressings of gray brick and Ningpo freestone. Unlike a good many red-brick edifices, the new police station is not finished off at the back or sides with cheap material such as common brick or plaster, but has red brick on all sides," the report said.

According to the report, the walls were covered with gray stucco, pleasing to the eyes. The doors and balustrade were of teak and properly strong and solid. In one or two places, such as the woodwork over a doorway in the inspector's quarters and the newels of the balustrades, there was carving highly to the credit of the Chinese artisan. A feature of the staircase was an ornamental arcade on the second floor, having three arches on brick columns with well-carved capitals.

Today, the building still looks a lot like what the 19th-century report described. Covering an area of 936 square meters, the building's front has a commanding ap-

pearance and each of the three stories is designed with a verandah in front.

It was not only a beautiful building, but also was functional as a 19th-century police station to house staff of different nationalities.

According to the North China Herald, the ground floor housed the foreign constables' room, a pantry, the charge room, the lost property office, the canteen, the billiard-room for foreign constables and the native constables' room.

The first floor had been divided into two unequal parts, one wing and part of the central space being devoted to the inspector in charge and the rest of this floor to Sikhs. The inspector had a front bedroom, and in the wing a sitting room, a dining room, a bathroom, pantry and cupboard. In the Sikh section was a sergeants' bedroom, a "joss" room, an interpreters' room and a large bedroom for the accommodation of Sikh constables.

On the second floor, which was for the European sergeants and constables, the whole central space was devoted to a bedroom and lavatory for the use of 12 constables. In one wing was a

bedroom for four sergeants with a bathroom, in the other a reading room with bathroom.

From this floor, access was gained to the tower, from the top of which a splendid view over the settlement was to be had. A lookout could communicate below with a speaking-tube. It's a pity that the ornamented tower has been demolished.

The police team was rather international from today's point of view. European sergeants and constables were in charge of management. Chinese constables were good at gathering information and managed Chinese-related cases. Sikh constables were responsible for guiding traffic while Japanese constables covered Hongkou and Putuo areas where many Japanese factories were located.

The Louza Police Station was designed to house up to 30 Chinese prisoners, 10 foreign prisoners and some 20 beggar. s As a big police station for the city's central areas, the building witnessed many historical events, the most noteworthy being the May 30th Movement.

On the morning of May 30, 1925, Shanghai police held 15 student protesters in the Louza Police Station. A huge crowd of Chi-

nese amassed outside demanded they be released but the police opened fire.

The shootings that killed nine and injured 14 sparked international censure and nationwide anti-foreign demonstrations and riots. The monument for the May 30th Movement now stands in the People's Park, a few minutes' walk from the red-brick building.

After 1949, the building became Pei Guang Middle School and since 1985 has been used by Shanghai Business & Tourism School, which is renowned for training future employees for the tourist industry including leading Chinese airlines. When the school's principal Li Xiaohua headed a renovation in 2009, he was as astonished as the visitors who saw the building first in the late 19th century.

"The old facade has been long covered by small white ceramic tiles so we treated it as only an ordinary old building. But when the workers removed some white tiles, the red bricks emerged and I was amazed by the breathtak-ingly beautiful facade. Then we made the decision to add an extra budget of 3 million yuan (US$450,000) for restoring historical details, including the facade, old wooden windows and an exquisite fireplace. The remaining 100 square meters of cells have also been preserved," Li says.

Today, the 19th-century police station is a signature building of the school. The ground floor serves as a mini museum with a Shanghai commerce theme which also displays the building's architectural history. The upper floors are used for conferences and students' social activities. In addition to tourism and commerce, the school also has a heritage protection major.

"We will host concerts in this heritage building. We also have a plan to let students to do historical research and design introducing boards of this building's history. Through this way, the value of this unique heritage can be passed on to the younger generation," principal Li says.

Yesterday: Louza Police Station   **Today:** Shanghai Business & Tourism School
Address: 101 Guizhou Rd
Architect: Arthur Dallas   **Architectural style:** Venetian Gothic style
Tips: The building is not open to the public but the facade can be admired from the school gate.

周六的晚上，宏伟的老闸巡捕房举行开幕仪式，新大楼就位于抛球场后面。开幕仪式是一个晚宴，随后在工部局巡捕房总督查麦克尤恩上尉主持下，来自欧洲的巡捕们发表演讲，并唱歌助兴。

三楼用作欧洲巡捕宿舍的大房间用旗帜和常绿植物装点着，布置了两张长桌，摆放为交叉的十字型。晚上8点以后，这里很快就聚集了130多人。他们跟随一位吹风笛的乐手，沿着蜿蜒的楼梯上来，为宴会拉开令人兴奋的序幕。

麦克尤恩上尉担任主持，他得到英国副领事卡尔斯先生、会审公廨法官和几位工部局董事的协助。参加晚宴的还有卡梅伦总捕、美国副总领事艾默斯先生、巡捕克鲁斯等人，他们坐在一张长桌的尽头。

参加晚宴的有来自艺术、科学、皇家海军、万国商团、海运和商业的各界代表，济济一堂。

由礼查饭店提供的晚餐扎实而丰盛，啤酒和葡萄酒的品质毋庸置疑。尽管出席者人数众多，暖气很充足，通风效果也很赞。没有人抱怨温度不适，也没有乌烟瘴气的感觉。

晚宴期间，吹笛者们在外廊游行，他们演奏的奇怪和弦活跃了气氛。如果某些演讲可以再精炼些，去掉一些瑕疵，演讲者和听众就不会有疲倦的迹象了。

歌曲演奏的水平远高于一般的餐后助兴，一个小提琴三重奏节目自然穿插在演讲中。当最后一位演讲者讲完坐下时已近凌晨1点，与会者对活动都感到很满意，在雾色中纷纷地回家了。

The new and magnificent Louza Police Station, behind the Racquet Court, was opened on Saturday evening with a dinner followed by speeches and songs, given by the European members of the municipal police under the presidency of the captain Superintendent, Mr J. P. McEuen, R.N.

In the large upper room on the third floor, which is intended to be used as a dormitory for the European constables, and which was decorated with flags and evergreens, two long tables were laid out with a cross table at one end, and here soon after eight more than a hundred and thirty convives assembled, preceded up the winding stairs by the piper, playing a stirring invitation to the banquet.

Captain McEuen was in the chair, and he was supported by Mr Carles, British Vice-Consul, the Mixed Court Magistrate, several members of the Municipal Council, etc. Chief Inspector Cameron, supported by Mr Emens, US Vice-Consul-General, was at the end of one of the long tables and Inspector Kluth of the other.

All the learned professions, the arts and sciences, the royal navy, the volunteers and the mercantile marine, and of course commerce were represented in the company.

The dinner, which was provided by the Astor House, was solid and substantial, the beer and porter as well as the wines were beyond questions, and notwithstanding the large attendance and the gas, the ventilation was so admirable that there were no complaints of the heat, and it was never impossible to see through the clouds of smoke.

The pipers marched up and down the verandah during dinner, enlivening the entertainment with their weird harmonies, and if some of the speeches might have been abbreviated without disadvantage, there were no apparent signs of weariness in speakers or hearers.

Songs, much above the average of after-dinner melody, and a violin trio had been liberally interspersed among the speeches and it was nearly 1am when the last speaker sat down, and the company, well satisfied, went home through the fog.

摘自 1890 年 1 月 3 日《北华捷报》
Excerpt from *the North-China Herald*, on January 3, 1890

诞生国歌的摩登剧场
Birthplace of China's National Anthem

简洁现代的黄浦剧院，是一座巧妙融合装饰艺术风格与中国元素的建筑。在这座中国建筑师设计的摩登剧场里，中华人民共和国国歌首次奏响。

剧场位于北京路贵州路的转角处，1934年开业时是金城大戏院，以放映电影为主，其摩登前卫的设计获得好评。

"大戏院的设计方案连贯、简洁、高贵、引人注目，没有冗余的细节。建筑既给人以亲切自然的感觉，造价又很经济。设计师没有专门为了视觉效果而设计，而是同时考虑功能与美观。建筑内外大量使用了直线条。"1934年英文《大陆报》关于金城大戏院的开幕报道写道。

大戏院的设计工作由华盖事务所完成。事务所由三位美国宾夕法尼亚大学建筑系毕业的中国建筑师——陈植、赵深和童寯创办。童寯之孙、同济大学教授童明认为金城大戏院是一个有挑战性的项目。

"这是一个不规则的转角基地，剧场空间就这么大，不可能设计一个'宫殿'，设计师还要面对入口和城市关系的问题。"他说。

2018年8月，童明教授在上海当代艺术博物馆策划了关于宾大回国的中国建筑师的展览。他研究发现，金城大戏院的设计手法非常考究，而如何在一个不太理想的环境中做出典雅的建筑，是这批中国建筑师在美国学习的基本功。

"金城大戏院这座建筑身上有很多现代主义的建筑元素，如整洁大气的立面，而必要的地方又有细微的装饰性元素，淡淡地体现了中国符号。这些都是这一批建筑师所特有的处理方式，例如外滩中国银行和南京路大新公司。他们把中国建筑的纹饰、花岗岩的建筑表面和剧院功能结合起来，不再是像罗马柱那样的古典风格，而是一个很现代、很城市的建筑。设计师对于建筑与城市基地、周边环境关系都有很好的考虑。"童明评价道。

1934年的报道提到，金城大戏院的外立面饰有5根用霓虹灯点亮的玻璃柱。戏院可容纳1800名观众，从第一排到最后一排座

位的观影效果都很好，每个座椅都安装了软垫，以确保观众舒适的感受。

此外，戏院的音响质量让观众身临其境。金城大戏院由中国商人投资兴建，开业后以放映国产电影为主，开幕影片是阮玲玉主演的《生活》。

1935年5月24日，电影《风云儿女》首映，日后成为中华人民共和国国歌的主题曲《义勇军进行曲》首次在剧场奏响。不幸的是，数月后创作国歌的年轻作曲家聂耳在日本游泳溺水去世，年仅23岁，他的悼念仪式就在金城大戏院举办。

1957年，大戏院成为上海淮剧团的演出场所，周恩来总理亲笔为剧场题写的新名——"黄浦剧场"沿用至今。20世纪80年代电影市场不景气，剧场部分空间不得不对外出租，先后开过音乐茶座和五金商场。

2016年，黄浦剧场历经修缮后，既恢复了历史旧貌，又更新了内部功能。老剧场被分为两部分——一楼的"黑匣子"小剧场和二楼的中剧场，以上演小型话剧、音乐剧以及儿童剧为主。

同济大学郑时龄院士曾专门在一次公众讲座上介绍，上海的很多优秀历史建筑其实是这些近代中国第一代建筑师的作品，如杨锡镠设计的百乐门、范文照设计的美琪大戏院、范文照与赵深设计的南京大戏院（今上海音乐厅）等。

他说，中国古代并没有"建筑师"的说法，只有"工匠"，"中国第一代建筑师"的概念始于20世纪初首批留学海外的中国建筑师回国。他们执业时正逢"东方巴黎"上海大规模建设的"黄金年代"。这批拿着洋文凭的建筑师或加入外国设计事务所，或像华盖事务所一样自己创业。

郑院士认为，与邬达克等已经成为"网红"的外国建筑师相比，这些近代中国建筑师的公众知晓度不够。设计金城大戏院的华盖事务所是一家重要的中国设计事务所，其三位合伙建筑师都很优秀。而华盖的作品从来不说明具体是哪位建筑师设计的，总是以华盖的名义作为设计师。除

了金城大戏院，华盖事务所的上海代表作还有恒利银行、大上海大戏院、梅谷公寓、浙江兴业银行和浙江第一商业银行等。

童明教授的展览向公众全面介绍了这一代中国建筑师的故事。

"他们刚回国时的作品都是古典主义的，但后来都转向对现代风格的探索。20世纪30年代，他们的设计和欧洲现代建筑差别不大，跟国际接轨是非常紧密的，思想上也极其先进开放，这一点从他们的学术论文中可以看到。他们做出的这些探索并非偶然，因为20世纪初是全球现代社会的发端。当时，经济产业与人类的生活方式都在变革，上海也进入现代都市的发展阶段，工商贸易蓬勃发展，有大量的城市建设，我们今天看到那么多的建筑作品都来自那个时代。"他说。

童明教授精心策划的展览观者如潮，展览的亮点既有他祖父参与创办的华盖事务所，也有简洁摩登的黄浦剧场。

昨天： 金城大戏院　**今天：** 黄浦剧场　**地址：** 北京东路 780 号　**建筑师：** 华盖事务所
参观指南： 黄浦剧场对公众开放，建议欣赏简洁现代的立面和室内中国元素的装饰细节。

Huangpu Theater, a modern design with Chinese elements, was the product of China's first generation of architects. It opened in 1934 as the Lyric Theater and was where the Chinese national anthem was first played as the theme song of a Chinese movie.

On the corner of Beijing and Guizhou roads, the theater was a motion picture house acclaimed for its ultra-modern design at the time.

The China Press, after it opened in February 1934, reported: "The entire plan of the theater was one of consistency, simplicity and dignity, giving a most striking effect and resorting to no sham features and meaningless expense to obtain result. The building as a consequence, appears intimate, natural and at the same time economical. Nothing was done for effect alone, utility being considered as well as effect. Straight lines are used throughout."

It was the work of Allied Architects, a firm of three Chinese architects, all of whom had studied architecture at Pennsylvania University in the United States.

"It was a challenging project," says Tongji University professor Tong Ming, whose grandfather Tong Jun was one of the three. "It was impossible to design a symmetrical building on the irregular site at a street corner, or design a 'palace' on the limited space. The architects also had to treat the relationship between the entrance and its urban context."

Tong Ming, who has curated an exhibition about the architects who returned from Pennsylvania University, says the Chinese architects learned how to design an elegant building on an imperfect site during their studies in the US.

"Modern architectural elements are everywhere inside and out of the Lyric Theater, from the neat, grand facade to decorative details with Chinese symbols. This was a unique style created by this generation of Chinese architects who merged Chinese elements with granite, monumental facades and new functions of a building. As a result, the Lyric Theater was no longer a classic building with Roman orders, but a very modern, urban building which had been taken into consideration with the urban site and its surrounding environment," he says.

Similar methods were employed on a galaxy of buildings at the time, such as the Bank of China on the Bund and the Sun Co. building on Nanjing Road E..

According to *the China Press* report, the facade of the Lyric Theater was made of five glass columns and illuminated by neon

lights. The theater was capable of accommodating 1,800 people. Every seat from the first row in the stalls to the last in the balcony commanded a perfect view of the screen. Each seat was upholstered and placed so that cramping was avoided and freedom of movement assured.

The quality of the sound equipment was so remarkable, the report said, that it gave the audience a life-like feel.

Invested by Chinese merchants, the Lyric Theater specialized in showing Chinese motion pictures. The inaugural program was "Life" starring actress Ruan Lingyu.

On May 24, 1935, "Children of the Clouds" had its premiere at the theater and its theme song, "March of the Volunteers," later became the Chinese national anthem. Sadly, the talented young composer Nie Er died only months later in an accident in Japan at the age of 23. His memo-

rial service was held in the Lyric Theater.

After 1949, the theater specialized in staging Huaiju Opera, a local opera originated in Jiangsu Province that was popular in Shanghai at the time. In 1957 the Lyric Theater was renamed Huangpu Theater by then Premier Zhou Enlai.

The theater underwent a renovation in 2016 which preserved the facade and historical details but divided the original auditorium into two smaller theaters to suit new performances such as small-scale dramas, musicals, children's dramas and Western contemporary dramas.

In a lecture on modern Chinese architects, another Tongji University professor, Zheng Shiling, noted that many of Shanghai's historic buildings were the works of China's remarkable first generation of architects.

He notes that traditionally the profession of "architect" did not exist in China, but only craftsman. However, things changed after the first generation of Chinese who studied architecture overseas returned after 1910, and a building boom started after that. The city was undergoing massive architectural changes as more peo-

ple poured into the "Paris of the East", and places for living, working and entertainment were built at an amazing speed.

When Chinese students returned with overseas architectural degrees, they joined foreign architectural enterprises or later opened their own companies.

"The work of Chinese architects has been undervalued and lesser known to the public compared with foreign architects like Park Hotel designer Laszlo Hudec," says Zheng who serves as city government's top expert for architectural preservation.

"Among them, Allied Architects was a very important archi-

tectural firm in modern Shanghai. Their works, including the Lyric Theater, the Metropole Theater of Shanghai and the Mercantile Bank, were all very modern. They focused on well-proportioned modern design," he says.

In the summer of 2018, Tong's exhibition at PSA museum displayed the life and work of these lesser-known Chinese architects.

"Some of them created classic buildings when they just returned from overseas but later they all switched to modern style. From their works and academic papers during the 1930s, I found they were keeping up with international trends and were extremely advanced and open-minded. Their pursuit of modernity was not occasional," Tong says.

"It occurred early last century when the whole world was transforming to modern society and modern metropolitan. That's why so many building stories were happening in Shanghai during that era," he says.

The exhibition entitled "The rise of modernity" featured the Allied Architects and the simple-cut Huangpu Theater.

Yesterday: Lyric Theater   **Today:** Huangpu Theater   **Address:** 780 Beijing Road E.
Architects: Allied Architects,
Tips: The theater is open to the public. Admire the facade and the Chinese decorative details inside.

白色酒店的文化体验

Cultural experience in a metropolis

白色的扬子饭店很特别，虽是西式酒店，却由中国人投资、设计、建造并且管理，在20世纪30年代并不多见。如今，白色酒店仍保留1933年开业时装饰艺术风格的立面，低调地位于红砖教堂慕尔堂的旁边。

上海金融史专家邢建榕研究发现，这家酒店与众不同。"老上海的华资酒店档次偏低，仅满足住宿功能，人员混杂。而扬子酒店的创办人曾在伦敦、巴黎等地经营餐饮业，他不仅从国外引进厨师和乐队，而且对团队进行专业严格的管理。"他说。

1933年扬子饭店开业时，被英文《大陆报》誉为"上海最新的摩登酒店"。这家"耗资百万美金建成"的酒店高达8层，由华人企业家C. F. Chong创办。他曾在欧洲成功经营餐厅酒店业，所以用中西合璧的理念来管理扬子饭店，分别从伦敦和巴黎请来西餐主厨，从夏威夷引进乐队在下沉式舞厅为客人表演助兴。

《大陆报》评论，酒店的设计"倾向于不同寻常"。这是"一座悦目的建筑，无论是住店客人，还是偶然光临餐厅和舞厅的客人，他们所需要的舒适服务都得到精心安排。"

值得一提的是，这家摩登酒店由一位中国设计师设计，主要建材的供货商也是中国人。

"这座位于汉口路和云南路转角的新楼从周边建筑中脱颖而出。虽然酒店所在区域正经历一轮快速的建设，但在未来很多年里，酒店都将是这里最与众不同的大楼之一……设计师李蟠（保罗）先生来自永年投资公司。他采用一种蓝色和黄色混凝土的色彩方案，为大楼营造了一种奇异的感觉。这种蓝色是明亮的，而

255

黄色则很深沉。"1933年10月19日的这篇报道写道。

除了色彩方案，扬子饭店的照明也十分精巧，外廊都被从下方点亮了。而酒店的室内照明也是一大特色，每个房间和走道里都装有用现代手法设计的吊灯。走道里的吊灯使用一种暗褐色的图案装饰，很有品位。

酒店的两个入口分别位于云南路和汉口路，用墨绿色大理石制成，为这座建筑增添了魅力。门口、大堂、部分舞厅和餐厅饰有一种深绿色大理石，硬木地板由一家中国地板公司铺设。这家酒店不仅由华人投资，用国产建材建造，经营团队也是经验丰富的华人经理。

同济大学钱宗灏教授指出，扬子饭店有很多折线形的图案，顶部层层收进，呈现了当时正风行的晚期装饰艺术风格。

这是一家时髦的酒店。《大陆报》报道提到，餐厅的装饰图案源于国外流行的中餐厅装饰风格，此外还安装了跳舞用的弹簧地板。餐厅的装修风格如此精巧而复杂，很显然管理团队致力于将这里打造成全上海最好的餐厅。

而酒店的220间客房都是西式的，配有最现代风格的家具和现代化设施，更重要的是还有阳台。有几间客房视野很好，能眺望上海跑马场，顶部楼层可以鸟瞰上海的城市风景。

邢建榕认为，扬子饭店不是国际饭店那种讲排场和档次的酒店，但又不同于早期低档而功能单一的酒店。

"扬子饭店注重居住的舒适性，餐饮有特色，又有丰富的娱乐与独特的文化氛围，给顾客一种'大都市的文化体验'。但它的价格又是可负担的，所以我很关注这家酒店。"邢建榕说。

在上海档案馆藏资料中，他找到一篇1939年研究扬子饭店的毕业论文，由沪江大学商业管理学系学生关宝定撰写。论文提到，饭店创办人"在附近南京路之处，向慕尔堂租地两亩七分余，由潘荣记承其建筑，费时逾岁，此巍峨之建筑物，方告落成，于民国二十二年十一月六日开始营业"。

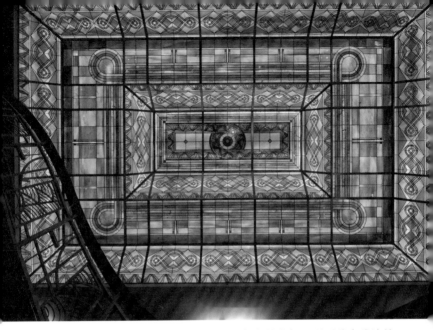

在论文的描述中，扬子饭店是一家设计和管理俱佳的现代酒店，有水汀装置加热，冬日里非常舒适。酒吧里既提供本地啤酒，也有从南京路先施百货订购的外国烈酒，夏季还供应冰激凌。酒店客房的卫生洁具和瓷砖都是从美国定制的，用精致的墙纸装饰，不同朝向房间设计有不同颜色花样的墙纸。

"房间顶上，都粉白色，墙壁之四周，表以花纸。花纸式样多种，朝南的房间，则颜色与花样，可以要稍为鲜艳的，朝北及向内的房间，则要稍为雅淡的；此为夏日之色彩配置，因雅淡之颜色，看了稍为阴凉悦目。"论文写道。

此外，这篇论文还特别介绍了客房的家具。"房间内之傢俬，由上海毛全泰木器行承做，二百余房间之内，没有一套傢俬之款式是相同的。其傢俬之价值，亦按房间之租价大小而布置之，其中有些贵的房间之傢俬，新式华丽，胜过富人之住宅的布置。"

1933年11月扬子饭店开业，

最初的四个月里生意火爆，保持满客状态，但次年因两家竞争对手——新亚饭店和国际饭店相继开业，生意开始下滑，后来又有所提升。

"价钱经济而设备完整的旅馆，扬子饭店也可算适合其条件了；例如印备信笺信纸，睡衣，拖鞋，毛巾及其他零用物品，以供旅客之应用，又如自来水之供给，公司方面，自备井水，以供旅客之无限使用。更有各式之简便机器之装置，其成本虽重，而工作能快利便，节省人工，且工作又能一律。其建筑之设计，极为周到，房租达四元以上一天的，则有私用的浴间与电话之安置，此外公司公用之浴间与电话，亦便捷妥当。"关宝定在论文中总结。

关氏落笔80多年后的今天，扬子饭店已经成为衡山集团的一家精品酒店，特色是怀旧音乐和玫瑰元素。在扬子饭店的"黄金时代"，著名的上海老歌《玫瑰，玫瑰，我爱你》曾在酒店舞厅上演。由于历经多次改造，室内历史细节留存不多。但不知为什么，许多年前由一个华人团队精心营造的文化气息仍荡漾在白色酒店里，宛若玫瑰老歌的余韵，久久不散。

昨天：扬子饭店　**今天：**扬子精品酒店　**建造年代：**1933 年
设计师：李蟠　**建筑风格：**装饰艺术风格
参观指南：酒店对公众开放。请欣赏建筑别致的外立面，入口处有关于酒店建筑历史的展览。

The Yangtze Boutique Hotel at the People's Square still retains the white Art Deco facade it had when it opened as the Yangsze Hotel in 1933.

"It's a unique Chinese hotel," says Shanghai historian Xin Jian-rong of the hotel which was invested, designed, constructed and managed all by Chinese.

"Unlike well-facilitated foreign hotels, the city's early Chinese hotels were mostly low-class and offered only accommodation. But

Yangsze Hotel was a different story, providing modern comfort as well as good cuisine, characteristic entertainment and a unique ambiance. What it offered was a cultural experience in a metropolis," says Xin.

When the hotel threw open its doors in 1933, it was called "Shanghai's newest modern hotel" and "a million-dollar establishment" by *the China Press.*

Standing next to the Moore Memorial Church, the eight-story structure is still a stylish boutique hotel. Nearly 85 years ago, it was founded by C.F. Chong, a veteran Chinese hotelier and restaurateur with many years' experience in operating restaurants in Europe. He incorporated both foreign and Chinese ideas in the hotel and entertained his clients with chefs from Paris and London and a Hawaiian band playing in the sunken ballroom.

The China Press said the hotel's design "inclined toward the unusual," which is "a pleasing building to the eye and well arranged for both comfort of those living within its walls and the casual customers who come to patronize its dining and ballroom".

"Situated at the corner of Hankow and Yunnan roads, the new structure predominates over all its surroundings and although that area is rapidly building up, it should be one of the distinctive buildings of the district for many years to come ... Mr Paul Li-pan, of the Mutual Investment Company, who designed the building, placed a touch of bizarre in the structure by working out a color scheme of blue and yellow concrete finish. The blue is bright while the yellow is very deep in color," said the report on October 19, 1933.

In addition to color schemes, elaborate plans were made for the illumination of the structure at sight as every verandah was illuminated underneath.

The hotel's two main entrances, one on Yunnan Road and the other on Hankou Road, were crafted with a blackish-green marble which added to the charm of the building.

The entrance, lobby and a portion of the ballroom and dining room were finished in marble of a darkish green color by the San Hai Marble Company.

Hardwood floors were laid by Y.S. Lee of the China Floor Company. The hotel was invested and designed by Chinese, built with Chinese suppliers and managed

by Chinese veterans in the restaurant and catering business such as T. C. Chang, formerly manager of the China Hotel.

"The facade features zigzag patterns and is shrinking layer upon layer until the top. The building displays Art Deco style of the later period inside out," says Tongji University professor Qian Zonghao.

It's a chic hotel inside and out. According to *The China Press*, the dining room was "patterned after the dining rooms of some of the most fashionable Chinese restaurants found in foreign countries and had a springboard type of dance floor. Decorations in the dining room are being carried out on an elaborate scale, the management apparently intending to make it one of the best if not the best dining room of this kind in the city".

Another feature was the interior lighting arrangement. Hanging lamps in the modernistic manner were fitted in every room and in the hallways, the latter being tastefully decorated with a dun color supplying the principal motif.

All 220 rooms in the hotel were in foreign style, furnished with modernistic furniture, equipped with all modern conveniences and, more importantly, with outside exposure and balconies. Several rooms provided a good view of the racecourse while the top floors offered an excellent panorama of Shanghai.

"Such a nice hotel charged only a moderate price but offered a cozy experience, which differed from the later big hotels like the Park Hotel that stressed on extravagance and showoffs," says Xin.

An in-depth synthesis of the Yangtse Hotel in 1939 that Xin scooped from shanghai archives backs his view. The paper was written by Guan Baoding, a business and management student from the then University of Shanghai.

According to this 1939 study, it was a well-designed, well-managed hotel of modern comforts.

The hotel bars offered local beer, famous foreign liquor provided by Sincere Co. on Nanjing Road and ice-cream in summer. The rooms were delicately furnished with wallpaper and bathroom facilities imported from the US. Rooms facing south were decorated in more bright colors while northern rooms appeared to be more light and elegant. Crafted by a local wood firm, not a single set of furniture of the 200 rooms appeared to be the same.

The hotel was fully booked during its first four months from November 1933 to February 1934. During the 1934 summer, the business declined due to the openings of two competitors, the New Asia Hotel and the Park Hotel but afterwards rose again.

"It was a moderately priced, fully facilitated hotel that provided even letter papers, pajamas, slippers, towels etc. The architectural design was very considerate. With a rate of US$4 a day, the customer can enjoy a room with a private bathroom and a private telephone," Guan concluded in his study.

It's been 85 years and many architectural details have been lost during renovations. But the hotel exhibits its past on the entrance and hosted the city's heritage day events.

In addition, nostalgic music and the rose theme, derived from the famous old Shanghai song "Rose, Rose I Love You" formerly staged in the ballroom, are highlighted in the hotel rooms and afternoon tea. Here and there, it seems the ambiance carefully created by a team of Chinese is still lingering in this unique hotel.

Yesterday: Yangtse Hotel   **Today:** The Yangtze Boutique Hotel
Date of construction: 1933
Architect: Li Pan   **Architectural style:** Art Deco
Tips: The hotel is open to the public. Please admire its unique facade and exhibition of hotel history.

36年前，当C. F. Chong先生挥别故乡广东到海外淘金时，他可能没想到会成为欧洲最好的中餐厅经营者之一。下月初，他担任经理的扬子饭店将在汉口路和云南路转角处开幕。

很可能，他并没有想到会进入酒店餐饮行业。到欧洲发展数年后，他拥有了多家中餐厅，既满足寻找新口味的外国顾客的需要，同时让旅居国外的同胞得以品尝家乡美食。

如今，Chong先生管理5家欧洲的大型中国餐厅，其中有两家最为知名——一家在伦敦皮卡迪利广场，另一家在巴黎医学院街上。

这些餐厅都以美食闻名，而Chong先生的名气也很大。皮卡迪利广场的那家餐厅规模很大，中外宾客如云，在伦敦家喻户晓。

两年前回国后，他考虑将商业和娱乐结合，决定在上海开设一家类似的饭店。这家最新的扬子饭店是他努力的结果，酒店将会复制他欧洲饭店的模式。中式美食将由专业的中国厨师指导烹饪，但美食爱好者在这里也能发现西式餐点。饭店从巴黎请了一位法国厨师，从伦敦请来一位英国厨师，以满足上海客人们的胃口。

新酒店的厨房是现代化的，配有最新的设备。这些设备根据最高的卫生标准设计，而在餐厅里也有烹饪空间。

Chong先生不是一般的餐饮业经营者。他在欧洲的职业经历很不一般，在英国非常有名，而英国也曾表彰他的贡献。

1925年英国博览会上，Chong先生作为一位"为许多国家提供美食的大师"被授予荣誉，表彰他"为英国的艺术、科学和工业的发展做出了贡献"。

When C.F. Chong, presently manager of the new Yangtsze Hotel situated at the corner of Hankow and Yunnan roads and scheduled to open about first of next month, left his native home in Canton some 36 or more years ago to go abroad and seek his fortune, he had little idea that he would become one of the outstanding Chinese restaurant proprietors in Europe.

It is possible that he had little idea that he would enter the restaurant and hotel business. Yet after a few years in Europe, he became the owner of a number of Chinese restaurants, designed to cater to the jaded appetites of foreigners in search of something new in the way of food and at the same time to cater to the tastes of his countrymen living abroad seeking orbits from their native land.

Today, Mr Chong is the head of five large Chinese restaurants in Europe, two of the best known being the Chinese Restaurant in Piccadilly

Circus and the famous Restaurant Chinois, 2 Rue de l'Ecole de Medecine in Paris.

Both of these establishments are famed for their cuisine and to anyone who has been in them, the reputation of Mr Chong as a restaurant man must speak for itself. The establishment in Piccadilly Circus is a large place which caters to both Chinese and foreigners and is widely known throughout London.

Returning to China two years ago, Mr Chong mixed business with pleasure and decided to open a similar establishment in Shanghai. The new Yangtsze Hotel has been the result of his endeavor and the establishment will be patterned after his places in Europe.

While Chinese food cooked under the guidance of expert Chinese chef will be served, the gourmet will be able to find dishes and viands of western countries at his disposal.

A French chef is being brought from Paris and an English chef from London to cater to the appetites of Shanghai patrons.

In regard to the kitchens of the new hotel, these are modern and furnished with all latest equipment. They are designed to attain the utmost in Sanitation while a cooking system has been installed in the dining room.

Mr Chong is not just an ordinary restaurant man. He has had an unusual career in Europe and is one of the best known in England. And England has honored him.

At the British Empire Exhibition in 1925, Mr Chong was honored for his achievement as a maestro in the serving of foods of many nations. In this connect, the citation states "For contributing to the development of the Arts, Sciences and Industries of the British Empire".

摘自 1933 年 10 月 19 日《大陆报》
Excerpt from *the China Press,* on October 19, 1933

楔型工人乐园

The grand hotel that became a palace for workers

西藏路120号是一座楔型建筑，它的建筑历史有趣地一分为二。1929年初建时是地段显赫的东方饭店，1950年又变成上海工人文化宫，成为一座"工人的学校和乐园"。

东方饭店由几位经营旅馆的华商联合投资兴建。饭店建筑于一块楔型的基地，由广东路、云南路和北海路围合而成。1926年项目启动前，原址是一家中式茶楼。

业主邀请新瑞和洋行（Davies & Brooke）设计酒店。这是一家老牌的外国设计事务所，代表作有兰心剧院和外滩礼查饭店。虽然东方饭店还不是新瑞和的经典之作，但酒店立面的形态与装饰契合了当时的审美潮流。巧妙的设计方案既充分利用了基地，又适应未来酒店运营者的需求。

"东方饭店由于建造在不规则的基地上，故建筑平面呈楔形。其西立面因为正好朝向跑马场，建成后成为视觉焦点，也成了建筑师刻意表现其才华的重点。"同济大学钱宗灏教授说。

他解释道，建筑的主立面正好位于楔形的端部，呈现在路人眼前的是梯形的三个面。建筑师巧妙地将其处理成三幅完全相同的构图：一至二层为基座，配以巨大的拱券门洞，三至五层作爱奥尼克柱式构图，上面设置一道腰线作为到六层的过渡。

这座大楼的另一个设计特色是阳台——沿广东路和北海路的立面有近百个一模一样的铸铁花栅阳台。

"建筑师成功应用了'模数'，其所形成的协调比例不仅让立面美观，也实现了一座酒店建筑的功能。"钱教授说道。

酒店一楼曾有气派的门厅和一家宽敞的餐厅。楼上是300间大小不同的酒店客房，房型丰富，从进深不到4米的单间到带阳台浴室的套间都有。建筑师还设计了两个内院，以便为部分房间引入光线和新鲜空气。也许为了充分利用昂贵的基地，大楼沿广东路和云南路的两侧开设了许多小商铺。

1929年东方饭店落成，1930年开业后虽面临周边酒店的激烈

竞争，但因为价格适中、环境清洁，颇受华商青睐，生意一直不错。东方饭店和同在人民广场的扬子饭店都是华人创办的酒店，但特色不同。扬子饭店以音乐表演和精致餐饮著称，东方饭店则首创将传统评书引入酒店。为了吸引更多客人，酒店在一楼东方书场提供传统戏曲和评书等文化娱乐活动。书场有400个座位，有一个带霓虹灯和麦克风的现代舞台。酒店甚至开设了一家无线电台来播放书场的演出节目。

研究这座大楼的建筑师倪正心认为，"这座大楼映射了中西文化的相遇和融合，曾经传统的中国演出在这座西式建筑里每天都在上演。"

1949年后，昔日上海跑马场变为人民广场，其周边建筑都陆续改变了功能和名字。东方饭店由上海市总工会接管，1950年10月1日重新开放为上海市工人文化宫。时任上海市长陈毅不仅出席了开幕仪式，还亲自为文化宫书写了一块匾额。如今，红色匾额上他书写的金字犹存——"工人的学校和乐园"。

倪正心认为，时代变迁让酒店变身为工人文化宫，她将这个案例比作荷兰建筑师库哈斯笔下的纽约下城俱乐部。

在《癫狂的纽约》一书，库哈斯写到这幢位于曼哈顿南端的摩天楼。大楼从形式上与其他摩天楼无异，有着密布的整齐开间、大片玻璃和装饰艺术风格的装饰，但在这张表皮下却聚合了丰富而拥挤的内容——从壁球、台球、拳击、游泳、高尔夫到浴疗、餐饮和住宿等。不同的内容以不同的空间方式填满整个建筑，堪称曼哈顿"拥挤文化"的代表。

库哈斯用"聚合器"一词来形容这种建筑形式与内容分离的建筑，而倪正心认为人民广场的这座楔型大楼成为工人文化宫以后，也是一个上海版的"聚合器"。

20世纪80年代就到上海市工人文化宫工作的祝少华副主任介绍，工人文化宫的形式是学习苏联的模式，1949年前叫工人俱乐部，是工人们交流活动的地方。1949年后，文化宫开办了补

习班和文化娱乐项目，如乒乓、猜谜、文艺、沪剧等，全市各单位工人中的佼佼者可以来参加活动。他提到，上海市领导将工人文化宫选在这里，是考虑到人民广场是上海地标。文化宫还举办创作培训班，一些从工人中被发掘培养的写作人才创作了《迷雾深处》《红色康乃馨》《蓝色马蹄莲》等经典话剧。此外，文化宫的茉莉花艺术团也非常出名。

为了承载一座工人文化宫的诸多新功能，昔日东方饭店历经多次改造，而祝少华对老建筑的历史遗存印象深刻。

"东方饭店设计考究，大楼集中供暖，底下有隔热层，门窗都是栗色木质的，可惜已经被换成铝合金的。客房里是带蚊帐的黄铜床，还有带四个凳子的大理石桌子，都曾经作为旧家具处理给职工。地坪是磨光石子的，花纹精致，楼梯都有护墙板。每个客房不大，大多数没有卫生间，但都有阳台。"他回忆道。

祝小华还提到，20世纪90年代中后期由于工厂效益不好，职工文化活动经费减少，文化宫不得不出租场地给证券交易所和羽绒博览会，在此期间只能举办一些小型文化活动。

2013年，文化宫获总工会全额财政拨款，回归公益功能后获得"新生"。原来对外出租的一楼现在是图书馆和小剧场，楼上各层分别用作职工援助服务中心、展览厅、培训、艺术活动室和作家工作室。近年来，工人文化宫也"与时俱进"，为企业提供更时髦的茶道、瑜伽和心理咨询等课程。考虑到年轻职工的习惯，传统的猜灯谜活动还推出了手机版。

倪正心研究发现，楔型大楼原来作为酒店的一个个小空间格局，因为变成了工人文化宫要改为做大空间，新功能塞进了历史空间。这在上海也并非孤例，文化广场、大世界和江湾大上海计

划的一些建筑在1949年后都改变了功能。

"上海很多历史建筑保存下来，但面临时代变迁和功能置换。库哈斯在《癫狂的纽约》一书中也表达了相同的观点，纽约的很多建筑就是保留一层壳，把新功能塞进去。那么我们的城市是不是癫狂的上海？文物保护建筑如何慢慢转变以适应变化，这是一个值得研究的课题。"她说。

昨天：东方饭店　**今天：**上海市工人文化宫　**地址：**西藏中路 120 号
建筑师：新瑞和洋行 Davies & Brooke　**参观指南：**建筑每天早 9 点至晚 9 点开放。

Social changes in China have split the history of a wedge-shaped building at 120 Xizang Road in two. As the stylish Grand Hotel, it stood at a prominent location since 1929 and set a precedent of adding Chinese storytelling entertainment in a hotel.

In 1950, it reopened as the state-owned Shanghai Worker's Cultural Palace and became "a school and paradise for workers."

The Grand Hotel was invested by several Chinese merchants who had run small hotels in the neighborhood. Enclosed by today's Guangdong, Yunnan and Beihai roads, the wedge-shaped site was home to a Chinese teahouse before the project was planned in 1926.

Chinese investors commissioned Davies & Brooke, an experienced Western architectural firm whose signature works range from Lyceum Theater, Astor House to Great Northern Telegraph Building at No. 7 on the Bund. The Grand Hotel might not be the firm's most eye-catching work, but the facade form and decorations catered to the aesthetic trend of its times. The site has been fully utilized and the almost flawless layout facilitated later operators of the hotel.

"Built on an irregular site, the Western-style facade facing the Shanghai racecourse naturally became a visual focus where the architect had intended to demonstrate his talent," says Tongji University professor Qian Zonghao, an expert in Shanghai architectural history.

"The major facade that we see today is at the top of the wedge. The architect smartly treated the three sides of the trapezoid fronting pedestrians into three identical compositions — the ground two floors with gigantic arched gateway as the base, the middle three floors graced by Ionic Orders and a belt line to transit to the top part which was treated in a simple way," the professor says.

Qian notes a noteworthy feature of the hotel building — the up to 100 identical cast iron balconies on the facade, a rare design on a classic building which functions as a modern hotel.

"The architect's successful use of 'module' not only created a harmonious aesthetic effect, but also fulfilled its functional use," Qian explains.

The hotel featured a grand vestibule facing a spacious restaurant in the center. Probably to make full use of the expensive land, the

facades on Guangdong and Yunnan roads were sprinkled with an array of small shops. On the upper floors, the architect placed 300 hotel rooms of all sizes, ranging from luxury suites with balconies and bathrooms to single rooms with a depth of less than 4 meters.

After completion in 1929, the Grand Hotel threw its doors open in 1930 and faced fierce competition from neighboring hotels. However the hotel enjoyed good business among Chinese merchants with its affordable prices and clean environment.

To attract more Chinese clients, the hotel offered performances such as traditional Chinese operas and storytelling on the ground floor. The theater featured 400 seats, a modern stage with neon lighting and microphones and even a wireless radio station for broadcasting shows and performances.

Ni Zhengxin, a Tongji University graduate who has researched the building, found it reflected the meeting and mingling of Eastern and Western cultures in Shanghai. For some time, traditional Chinese performances were presented every day in this Western concrete architecture.

As Shanghai racecourse became People's Square in 1949, its surrounding buildings all changed to new names and new functions. The Grand Hotel was purchased by Shanghai Workers' Union and reopened as Shanghai Workers' Cultural Palace on October 1, 1950.

The then Shanghai Mayor Chen Yi not only attended its opening ceremony but also wrote a commemorative plaque for the palace. His calligraphy in gold still glistens on a scarlet wooden board — "a school and paradise for workers."

"Political changes have prompted new functions which required big spaces for library and training classes into the original layout featuring hundreds of small hotel rooms. Then the historical building became a new complex that mirrored the Manhattan skyscraper in architect Rem Koolhaas' book 'Delirious New York' , a structure stuffed with multiple functions including squash, billiards, boxing, swimming, golf, spa, restaurants and accommodation, a typical representative of Manhattan's culture of congestion," Ni comments.

According to Zhu Shaohua, deputy director of the palace, the form of workers' cultural palace derived from the workers' club, a place for the working class to communicate and enjoy activities before 1949. The palace hosted cultural classes and entertainment for Shanghai workers since its opening, such as table tennis, chess, music, dancing, calligraphy, painting and Huju Opera.

"The municipal government chose this building because it's located at the landmark People's Square. Only the most excellent workers had the opportunity to participate in activities here. For the past decades, we've discovered many talent from ordinary workers. We organized training classes or workshops to help them develop their interests and talent like writing novels, creating dramas or other artistic performances," says the director who has worked there since the 1980s.

To adapt to the new functions, the former hotel building has been renovated several times since the 1950s. But Zhu still has a profound memory of historical relics left from the era of the Grand Hotel.

"It was a well-designed, fully heated building. The doors and windows have chestnut-hued wooden frames which had unfortunately been replaced by al alloys. Hotel rooms were not big, most of which had no bathrooms. But many had balconies. The rooms are usually furnished with a copper bed cover, a marble table with a set of four stools and a dressing table with marble desktop. In public areas, the floors have exquisite patterns while the walls beside staircase were adorned by wooden dados," he recalls.

The palace endured a hard time since the 1990s when many state-owned factories saw a drastic decline in business and many workers were laid off. Without financial support from local factories, the cultural palace had to rent out space for commercial use, such as for a stock exchange agency and for an exposition of down products. Only small-scale cultural activities were hosted during "the difficult 20 years".

Since 2013 the cultural palace received more financial support from Shanghai Workers' Union and returned to its non-profit role. Now the ground floor, which was rent for commercial use, houses a library and a small theater. The upper floors serve as the workers' aid and service center, an exhibition hall, training, artistic activity room and writers' workshops.

Zhu's team also walked out of the cultural palace to send training and cultural activities into office buildings and big enterprises. To attract young workers and white-collars, they provided trendy lifestyle classes like baking, tea ceremony, yoga and psychological consulting. Even riddle-guessing games traditionally written on Chinese lanterns now have mobile phone versions.

Ni notes that the Grand Hotel building is not the only case that experienced a dramatic transition of functions during social and political changes. After 1949, the famous dog-racing course became Shanghai Culture Square, and many buildings constructed for official use during the 1929 Great Shanghai Plan, launched by the then Shanghai Municipal Council, all changed to new roles after

1949.

"Many of Shanghai's solid old buildings survived the times and confronted with change of functions and times. In Koolhaas' 'Delirious New York,' many New York buildings only had a shell preserved and were with many new functions. So is our city a delirious Shanghai? For a heritage building, how to gradually transform and adapt to the changes is a subject worthy of further studies," Ni says.

Yesterday: Grand Hotel   **Today:** Shanghai Workers' Cultural Palace
Address: 120 Xizang Rd M.   **Architects:** Davies & Brooke
Tips: The building is open to the public from 9am to 9pm daily.

两个男人带着各种抽鸦片的工具仓皇逃到酒店的屋顶，但他们速度没有快到逃脱法律的制裁，昨晚对虞洽卿路（今西藏路）东方饭店和远东饭店的突袭快如闪电。

探员们看到一眼两人的脚跟消失在一条通往东方饭店屋顶的狭窄楼梯。他们全速追赶，两个倒霉的人还来不及扔掉罪证，就被抓获了。

突袭在警察和万国军团俄国分队抵达一两分钟后就大规模展开了。两家酒店被一圈带着闪亮刺刀和钢盔的俄国志愿兵包围了。所有逃跑通道都被堵住，突袭的警力分为几组，他们带着手枪首先来到东方饭店搜查。

很快两名罪犯被带到门口，他们走过人群，中国男人和女人们惊讶地睁大了眼睛。人们呆呆地站在酒店大堂里，前面是后备队的印度锡克族警察、穿钢背心制服的警察和来自老闸巡捕房与后备队的探员们。

这两个男人就是在屋顶被抓获的两人，他们身后有许多抽鸦片所需的物品，还有天平和一些白色粉末。与此同时，突袭的几路人马走进酒店的每个房间，惊吓了整幢楼里很多打麻将的聚会。

当探员们把人们带到角落里搜查时，其他的警察认真在房间里检查，牙刷、床垫、枕头、毯子和床单都被无礼地扔到地上。衣柜里的东西也被拿出来仔细检查。

搜查人员连帽子和鞋子也不放过，他们戳戳椅子和垫子，直到确认没有罪证才满意。

突袭也有好笑的一面，不止一次有抗议的华人被拖下床，他们睡眼惺忪感到震惊，但还是屈从于中国探员习惯性的"劫掠"。

然后就是彻底搜查床铺，床垫被扔到一头，枕头和床单扔到另一头。原来睡在床上的房客现在已经清醒了，以敬畏的眼光看着，颤抖不已。

　　就连浴室也没有逃过检查。一个中国先生很不幸，他当时只有一点肥皂沫来遮住身体。微笑的探员们叫他站起来，然后认真检查浴缸里的热水。

　　很幸运，他们只找到了一些浴盐。不过，他们怀疑地看了看浴盐，直到一位外国官员解释这些确实是无害的。

Hurriedly scampering to the roof of the hotel with all the various paraphernalia used by opium smokers, two men were not quite quick enough to avoid the clutching arm of the law during two lightning raids which were carried out at the Grand and Far Eastern hotels on Yu Ya Ching Road (today's Xizang Road) last night.

Detectives, catching sight of their heels vanishing up a narrow stairway leading to the roof of the first-named hostelry, put on all speed, and the two luckless fellows were seized before they had time to dump the incriminating objects carried by them.

The two raids were carried out on a major scale, a few minutes after the arrival of the police and Russian Regiment S.V.C. vans.

The two hotels were surrounded by a ring of glinting bayonets carried by steel-helmeted Russian volunteers. With all avenues of escape blocked, the raiding police parties formed into several groups and, with pistols in hand, made their way into the Grand Hotel, which was the first to be searched.

Within a few minutes two prisoners were being led to the doorway through a crowd of wide-eyed Chinese men and women, who stood spellbound in the hotel lobby behind the rifles of Sikh members of the Reserve Unit and the steel-waistcoated uniformed police and detectives from Louza and the Reserve Unit.

These were the two men seized on the roof and, behind them, came the various objects necessary for opium smoking, with the addition of some scales and some white powder. Meanwhile, raiding parties were entering every room in the hotel surprising numerous mahjong parties throughout the building.

Herding the inmates into one corner while detectives searched them, other police literally went through the rooms with a tooth comb, mattresses, pillows, blankets and sheets being unceremoniously dumped on the floor while the contents of wardrobes were brought out for further scrutiny.

Even hats and shoes were searched and chairs and cushions prodded before the parties were satisfied that nothing incriminating was stored there.

The raids were not without an amusing side, on more than one occasion protesting Chinese being dragged out of bed, sleepy-eyed and somewhat shocked, to submit to the customary "running over" of the Chinese detectives.

Then would come a thorough searching of the bed with the mattress thrown to one end and the pillows and sheets at the other while the former occupant, now thoroughly awake, looked on with awe and trepidation.

Even bathrooms were not exempt from the searches as one unfortunate Chinese gentleman found to his cost, when, with nothing but a lather of soap to cover his nakedness, he was asked to stand up while smiling detectives diligently searched through the hot water in the tub.

Luckily nothing was found there, except some bath salts which were looked upon with some suspicion until a foreign officer explained that they were quite harmless.

摘自 1939 年 3 月 15 日《北华捷报》

Excerpt from *the North-China Herald* on March 15, 1939

中西合璧的格致书院

East meets West in the Utmost Pursuit of Knowledge

上海格致中学的草坪上有两座铜像，一中一西两位学者，并肩交谈。两座铜像与幸存的老校舍，默默讲述着学校极不寻常的历史。

两座雕像分别是中国科学家徐寿和英国翻译家傅兰雅，都是格致中学前身——格致书

院的创办人。格致书院创办于1876年，取义儒家理论中的"格物致知"，即"推究事物的原理，从而获得知识"。而学校的英文名"Chinese Polytechnic Institution"（中国理工学院）则源自1836年创办于英国伦敦摄政街的皇家理工学院。格致书

的创办者期望这所新式学堂成为"中国版的皇家理工学院"。

作为近代中国最早中西合办、培养科技人才的新型学堂，格致书院中西两个名字都名副其实。陈列着科学仪器的书院举办科学展览，开展科学演讲，进行科学普及，是晚清上海的科学活动中心。

格致书院由英国驻沪总领事麦华陀（Sir Walter Medhurst）倡议创办。他最初建议为华人建一间阅览室。这个提议得到中外各界人士的积极响应，最后阅览室扩展为一所传播西方科技知识的学校。

1876年6月，格致书院正式创办。它虽有书院之名，但并非传统书院，设在租界，却并非租界的公立学校，虽有传教士参与办学，又并非教会学校，有官方资金支持，但也不是官办学校。著名上海史专家熊月之认为，这所学校"不中不西，亦中亦西，非官非民，亦官亦民"，是中外教育史上一个罕见的机构，亦是上海这个特殊城市的产物。

"当时上海华杂洋处，存在着事实上不受中国政权控制的租界，居住着大批外国人。这些人中，有凶恶的侵略者，贪婪的冒险家，也有虔诚的宗教徒，认真的文化人，更多的是几种身份兼而有之者。不管出于什么动机，他们当中相当一些人希望上海了解西方科学技术。同时，上海汇集了一批有世界眼光，比较懂得西方科学技术的中国知识分子和绅商，如徐寿、华蘅芳、郑观应、唐廷枢。中外这两部分人的结合，促成了格致书院的产生。"他在论文《格致书院与西学传播》中写道。

格致书院盛大的开幕仪式中西合璧，来宾既有上海本地的军政要人，也有来自多个国家的外国领馆官员和商人。1876年6月24日的《北华捷报》报道，麦华陀在仪式中回顾了他三年前提议建立阅览室向华人介绍西学后，中外商界人士慷慨响应，他的朋友傅兰雅建议应该办一所学校。这个想法通过大家的努力终于实现了。

"创办学校的想法特别得到徐寿先生的热切支持。徐寿和他的儿子都参与到学校董事会工作中，他努力募集了近四分之三的经费。"麦华陀在开幕仪式上说道。这位英国总领事还提到，建校的计划在英国国内也传播得很快，许多人答应为学校寄来仪器。"而徐先生在家中收藏了很多科学仪器，这在华人里是不多见的。他答应将自己的藏品借给学校使用。"他说。

报道介绍，格致书院的校舍"平凡但实用"，一些欧洲公司提供了不少好看的仪器，如电报装置、火车头和马车的设计图、一对地球仪、一个可转动的太阳仪、一支气压表、一个大火车头模型和一张地图，用于"切实有效地将铁路介绍给中华帝国"。学校的图书馆最初藏书并不多，但有西方科学书籍的中译版，还有一套关于中国农业的书籍。

如今，"平凡"的19世纪校舍早已不存。在近代中国动荡的政治风云中，中西合办的格致书院历经多次变迁。书院从1915到1941成为工部局管理的格致公学，1942年到1945年是上海特别市市立格致中学，1945年到1949

年为市立格致中学，解放后则一直是格致中学，2005成为上海市首批"实验性示范性高中"。

同济大学郑时龄院士提到，1926年兴建的老校舍在2003年学校扩建时几乎被拆除了，因为老校友坚持保留了一部分。他回忆这座作为工部局学校而建的老楼虽然一般，但"现在留下的门头是最有特色的"。他曾与章明、张姿两位建筑师合作，为格致中学设计了新校舍。

1925年1月15日，英文《大陆报》刊登了这座幸存老楼的方案。"新大楼有18间大小的教室、一个艺术教室、一间大的科学教室、一个讲座教室和手工培训教室。大楼底层中央是一个礼堂，可以用于做体育馆使用。"报道写道。

格致中学文化研究室主任柯瑞逢提到，1927年落成的新校舍设施完备，落成当年招生报考者云集，学生猛增至278人。"从此格致公学扬帆起航。1929年后，学生每年都稳定在500人以上。现在的格致中学是在1876年开办的格致书院原址周边扩建的。后来历经改扩建，至今成为四条马路'保卫'的格致中学独立社区，占地约20亩，是全国百年名校中为数不多的校址不改、冠名一脉相承的学校。"他说。

如今，幸存的老校舍已被改造为校史馆。馆里展示了为创办这所学校而做出贡献的中外人士，包括徐寿和傅兰雅。这两位19世纪学者的雕像每天都在默默注视着格致中学的学子们。

"在19世纪，徐寿等一批中国人热衷于学习和介绍西方科学。他们先仿制西式机器，又翻译相关的西学论文和专著。最后，他们想要培养更多这方面的人才，所以创办了格致书院。"柯主任提到。

如今，位于上海市中心的格致中学成为第一家引进美国麻省理工学院FabLab实验室的中国学校。格致中学的毕业生中有13人成为中国工程院和中国科学院的院士。

"一个多世纪以来，学校坚持发扬'爱国''科学'的传统，走'科教兴国'道路，为国家和民族培养了大批人才，以优异的办学业绩享誉海内外。"老楼校史馆展览的前言写道。

昨天: 格致书院　**今天:** 格致中学　**地址:** 广西北路 66 号

建造时间: 1927 年（始创于 1876 年）

参观指南: 建筑不对外开放，可以从广西路欣赏老校舍精致的门头。

Statues of two 19th-century scholars--one British, one Chinese--stand shoulder to shoulder on the lawn of Shanghai Gezhi High School, a stone's throw away from People's Square. The statues and a surviving section of old school building tell an extraordinary story of this East-meets-West school which first taught Western sciences in China.

They are British translator John Fryer (1839-1928) and Chinese scientist, translator and mechanist Xu Shou (1818-84), two core founders of this unique school, which was managed by a committee of both Chinese and foreigners.

The school that opened in 1876 was named the Chinese Polytechnic Institution, hoping to emulate the well-known establishment of a similar name in London's Regent Street but on a smaller scale. Its Chinese name, Gezhi Academy, was inspired by "ge wu zhi zhi" (格物致知), from the Confucian theory meaning "study the underlying principle to acquire knowledge or pursue knowledge to the utmost".

As China's first new-style school to import education of natural sciences and even edit textbooks in the subject, the institution lived up to both Chinese and English names.

Sir Walter Medhurst, then British consul in Shanghai, was the original promoter of the scheme. At first, it was only intended to be a reading room for local people. Foreigners and Chinese subscribed funds and finally the scheme resolved itself into the form of a polytechnic institution where the arts and science of the West were taught.

Shanghai historian Xiong Yuezhi found it was a rare example in both Chinese and foreign educational history. The founding of the school was advocated by a foreigner, but it was not a missionary or expatriates' school. Chinese elites participated but it was not entirely managed by Chinese. Chinese officials exerted influence but it was not attached to the Chinese government. It's neither Chinese nor foreign, but both Chinese and foreign. It's neither official nor private, but combines both official and private forces. The school was the product of a unique city like Shanghai.

"Among the expatriates in old Shanghai were fierce invaders, greedy adventurers, devout religionists and serious cultural scholars, some of whom might have

multiple of these roles. However, some of them had the intention to introduce Western sciences and technologies to Shanghai. Meanwhile our city gathered a group of Chinese intellectuals and merchants like Xu Shou. The meeting of these foreigners and Chinese brought to the birth of this unique institution," he writes in a paper of this unique institution.

This special feature was fully showcased at the opening ceremony attended by "principal civil and military mandarins of the native city in numbers" and "a numerous attendance of consular officials of different nationalities, merchants and other residents."

According to a report in *the North-China Herald* on June 24, 1876, Medhurst recalled his original idea of founding a reading room to introduce Western arts and sciences to Chinese some three years before. It met a generous response from both foreign and Chinese merchants. His friend, Mr Fryer, suggested the proposed institution should be something more than a mere reading room, and that an endeavor should be made to form it into a polytechnic institution and school of art.

"That idea was taken up with earnestness and especially by Mr Hsu, who with his son had been added to this committee and through whom above three quarters of the funds have been obtained," Medhurst said during the opening ceremonial. "The proposition was also well circulated at home and many people there also promised to help the institution by sending specimens of their arts and manufactures. Mr Hsu is one of the very few Chinese who have made collections of scientific apparatus at their own homes. He has promised to lend us a large proportion of his collection for the use of the polytechnic."

In addition to honorable guests and the opening of the school building which is "plain but suitable for the purpose", many European firms provided beautiful specimens from telegraphic and electrical equipment to plans of locomotives and wagons. There were also a pair of globes, a movable orrery, a clock-barometer, a large model of a locomotive, and a sketch map prepared for "the feasibility and most effectual means of introducing railways into the Empire of China".

The surviving part of the 1927 building has been renovated into a museum of Gezhi High School, whose history dates to 1876 when it was founded as the Chinese Polytechnic Institution.

The school library, though not overstocked with books at the beginning, contained translations of foreign scientific books and a collection of works on Chinese agriculture.

Today, the 19th-century "plain building" has long gone and the institution underwent changes during political changes. It became the Polytechnic Public School run by Shanghai Municipal Council from 1915 to 1941; Gezhi Middle School of the Shanghai puppet government from 1942 to 1945; Shanghai Municipal Gezhi Middle School from 1945 to 1949 and Shanghai Gezhi High School from 1949 to the present day. A 1920s building has survived.

"This historical building was almost demolished during an expansion around 2003. Thanks to the efforts of some old alumni, a small portion of the building has been preserved. The building was 'plain' but the preserved entrance gate was exquisite," says Tongji University professor Zheng Shiling, who co-designed the school's new buildings with colleagues Zhang Ming and Zhang Zi.

On January 15, 1925, *the China Press* released plans for "the new building of the Polytechnic Public School for Chinese" approved at a meeting of the Chinese Education Committee.

"The plans call for a building containing 18 classrooms of varying sizes and an art room, a large science room, a lecture room and manual training rooms. The assembly hall is centrally placed on the ground floor and sufficiently removed from the classrooms to allow of its being used as a gymnasium also," the report said.

Ke Ruifeng, director of the school's Gezhi cultural research office, says: "The new building completed in 1927 was acclaimed 'adequately facilitated' and attracted more recruitments. Since then it seemed like a new start of our school. The number of students soared to 278 the same year and grew to be over 500 after 1929. Among China's galaxy of history-rich schools and universities, it was one of the few that still perches on the original location and luckily maintained the original Chinese name--Gezhi."

The surviving part of the 1927 building has been renovated into a school history museum, which exhibits photos of a galaxy of Chinese and foreigners who contributed to the forming of the institution, including John Fryer and Xu Shou, whose statues look over Chinese students every day.

"In the 19th century Xu and his fellow Chinese were enthusiastic to learn and introduce Western sciences. They first learned technologies by imitating Western machines, then translated related theories and books. Finally they wanted to train more talent in this field and that how our school was founded," Ke says.

Today the school is one of the city's top high schools in the heart of Shanghai. Gezhi imported MIT's first FabLab in a Chinese school and among their graduates are 13 academicians of the Chinese Academy of Sciences and Chinese Academy of Engineering.

"One hundred and thirty years or more, this school has adhered to the tradition of patriotism and science, advanced along the road of rejuvenating China by developing science and education," the opening introduction of the school museum notes.

Yesterday: Polytechnic Public School for Chinese(originally Chinese Polytechnic Institution in 1876) **Today:** Gezhi High School **Address:** 66 Guangxi Rd N.
Date of construction: 1927
Tips: The building is not open to the public. The exquisite gateway can be admired from Guangxi Road.

中国的科学进步
Chinese progress in science

昨天，一个重要集会在摄政街帝国理工学院举行。人们见证了关于一个新仪器展示的有趣的光学现象。这个仪器是为上海一所学院准备的。

来自理工学院的金先生 (J. L. King)负责演示。他先做了一个简短的演讲，解释了产生这种现象的科学原理。他还表示这会唤醒中国人对科学教育的重要性的意识。

这家中国的学院主要源于原英国驻沪总领事麦华陀的影响与努力。他担任学院上海委员会的主席，傅兰雅先生担任名誉秘书长。

他们的想法也得到我们驻华大使威妥玛（Sir Thomas Wade）和一些华人精英给力的支持，其中有已经在欧洲知名的李鸿章和上海道台冯浚光。此外还有徐寿和他的儿子。他们父子以科技工艺和科学启蒙闻名，致力于引进西学来为国服务。

新学堂的设计首先是为了满足爱国之情。在上海，一座大楼已经盖起，在那里举办了讲座，展示了有趣的仪器和实验过程，而提供合适书籍的阅览室已在很好地使用了。未来将举办一个中国国际博览会，将来自东方和西方最有趣的产品并列展示。此外，傅兰雅先生还在上海编辑了一本新的中文科学杂志。

金先生的讲座让观众非常满意，从各方面来说都很成功。将被寄往中国的仪器包括伯恩公司（Messrs. Bourne & Co.）的几个高速引擎。这些引擎有望成为未来的蒸汽引擎。

其中一个引擎将用来驱动一个金刚砂轮，运作起来就好像一个可以快速抛光金属和削尖切割工具的旋转锉子。这个机器吸引了很多关注。伯恩公司也寄出了一些很好的瓷器样品，大多是英国明顿公司的产品。

目前中国这个项目最有前途的一点，就是它由中国人自己采纳并提出的。所有这些创新的第一步都是最艰巨的，现在看来很可能取得成功。

Yesterday an influential meeting took place at the Polytechnic Institution, Regent Street, to witness some of the most interesting phenomena connected with the popularization of light and other branches of optics as shown by a new apparatus intended for an institution at Shanghai.

The demonstration were conducted by Mr. J. L. King of the Polytechnic who delivered a brief lecture explanatory of the scientific principles, which determine the phenomena and who stated that the occasion derived its chief interest from the fact that it marked an awakening in the Chinese mind to the importance of scientific instruction.

The Chinese Institution owns its origin mainly to the influence and exertions of Sir Walter Medhurst, lately our Consul in Shanghai who was Chairman of the Shanghai Committee and Mr. Fryer who acts as Honorary Secretary.

Their views have been energetically supported by our ambassador, Sir Thomas Wade and by many of the leading mandarins, among whom may be mentioned Li Hung-Chang, whose name has a European celebrity; Feng Chu-ju, Taotai of Shanghai, Hsu-Tsuch-Tsun and his sons, who are well known throughout their own country for their technical skill and scientific enlightenment; and by many others who favor the importation of Western knowledge of every kind as calculated to render the most important service to their country.

The design of the new institution is to make a first step towards satisfying this patriotic aspiration. A building has been erected at Shanghai, within which lectures are to be given and interesting apparatus and processes shown, and a reading room provided with suitable works is already in active operation. These measures are expected and designed to lead up to a Chinese International Exhibition at which the most interesting productions of the East and West will be brought into juxtaposition. A new Scientific Magazine, in the Chinese language, has been established in Shanghai edited by Mr. Fryer.

Mr. King's lecture gave great satisfaction to his visitors and was in every respect most successful. Among the articles being sent out to China by Messrs. Bourne & Co. are several of their new high speed engines, which are believed by the most competent judges to be destined to become the steam engines of the future.

One of these engines will be employed to drive an emery wheel, which acts like a rotating file in rapidly polishing metals and sharpening cutting instruments. This machinery attracted much attention. Messrs. Bourne & Co. are also sending out some fine specimens of porcelain, mostly of Minton's make.

The most promising feature in connection with the present movement in China is that it is adopted and presented on by Chinese themselves. The first step in all such innovations is the most arduous, and it appears now likely to be taken with success.

摘自 1877 年 7 月 2 日《北华捷报》
Excerpt from *the North-China Herald*, on July 2, 1877

浓缩欢笑的大世界

A condensed world of entertainment

"上海大世界是一个浓缩的世界"，著名建筑师邢同和说。童年时，他曾在大世界度过愉快的周末时光，半个多世纪后又负责修缮这座百年沧桑的建筑。

1917年营业的大世界是一家大型游乐场。创办人黄楚九曾经营中法大药房，制造销售"艾罗补脑液"获利致富。此后他跨界经营，1912年先创办新新舞台、楼外楼屋顶花园和新世界游乐场等娱乐项目，从新世界退股后又集资开办了大世界。

上海档案馆研究员张姚俊研究发现，黄楚九很会想办法迎合中国人的需求。"他利用中国人崇洋媚外、不信任国货的心理，臆造出一个洋博士'Dr. Yale'和他发明的'艾罗补脑液'。所谓的'补脑液'充其量只是保健品，却十分畅销。而大世界也深受欢迎，因为门票价格低廉，可以一直玩到关门，又有华人喜爱的娱乐节目，如杂技和多种方言的戏曲。"他说。

早期的大世界是一座形式简单的木建筑。根据1917年7月21日的《北华捷报》报道，游乐场有一个用于多种娱乐表演的露台剧场，此外还有很多房间，用于餐饮、货摊、台球室和其他演出。

"其中一大特色就是设有露天剧场，这与欧洲很多地方的月神公园（Luna Park）风格相似。在这里开心的中国人可以体验骑木马、空中走钢丝、游艺打靶等活动的刺激。此外，这里还有全景画、放电影、动物展览和摩天轮，让人想起英国的市集。"报道写道。

虽然面积不大，大世界游乐场还是被视为"中国的新奇事物"。游乐场从内到外都设计了电灯照明，夜幕降临时成为璀璨的地标。这里全天都有娱乐项目。露天剧场以杂技表演为主，还有10个小剧场上演多种地方戏曲节目。游乐场平均每天吸引1.5万人，节日达到2.5万人，其中大多数都是中下阶层的华人。

由于营业兴隆，1924年大世界翻新改建。钢筋混凝土结构的新楼由周惠南打样间设计，平面呈L形，沿西藏路延安路的东南转角布局。转角建有一座55米高的西方古典风格塔楼，成为上海大世界醒目的标志。

建筑内部的布局新颖精致，底层中央是露天剧场，周围有长达百米的天桥式扶梯。天桥蜿蜒交错，繁而不乱，上下贯通，南北相连。游客们既可以通行，又能凭栏观看露天演出。大世界的大厅里是12面著名的哈哈镜，剧场两侧的建筑设有小型剧场和舞厅，每层沿马路都有宽阔走廊供游人走动，还有一个屋顶花园。

大世界经营成功，黄楚九忍不住再次跨界，涉足金融业开办日夜银行。但是很不幸，这一次他投资失利，只好把大世界卖给青帮头目黄金荣经营。1931年，黄楚九病逝，年仅59岁。

1954年大世界由上海市文化馆接管，改名人民游乐场，"文革"中改为青年宫。1981年，大世界终于恢复原名，并增加了新的游乐设施。不过随着新型主题乐园的出现，大世界渐渐冷清，2004年关门停业，历经多年的空置和修缮。

20世纪40年代，还是个小学生的邢同和曾在大世界度过许多个愉快难忘的下午。

"我的爷爷是大世界旁边李鼎和笔庄的老员工，他周末下午经常带我去大世界玩。他在里面看戏时，我就自己去探索这座多层的游乐场。"他回忆道。

"我在带栏杆的楼梯跑上跑下，在哈哈镜前笑，到不同的剧场看表演。在我童年的记忆中，大世界好大啊。"邢同和说。

2004年，这位建筑师回到大世界做修缮前的勘察工作时，却发现大世界不仅没有童年记忆中那么大，而且破败不堪。

"栏杆断裂，钢筋裸露，地板卷翘，哈哈镜满是灰尘。在过去80多年来，大世界每天要服务上万人次，2004年时看起来就像是一个老化的仓库，我很伤心。"他说。

邢同和认为，大世界是一个浓缩了上海娱乐和文化生活的世界。在修缮工程启动前，曾有关于大世界未来定位的争论。人们讨论是把它打造为一个时尚的动漫主题乐园，还是保留传统。

"大世界从一开始就为普通百姓建造的，也给年轻艺术家们提供了一个表演和练习的平台，其中有些人后来成长为名角。大世界可

以改造成一个动漫乐园，但历史的味道将会永远消失，而动漫乐园在任何地方都可以建造。"建筑师说。

最终，邢同和他的团队根据大世界黄金时期的外观进行修复。他们查阅了历史照片和原始图纸，在考虑现代功能的同时，保护修复了马赛克地砖和回廊等不少历史细节，还在顶部重建了已拆除的凉亭。这个颇具挑战的项目历时十年，于2016年完成。

2017年3月31日，这座诞生于老上海的游乐场在百岁生日之际，修缮竣工重新对外开放。如今，大世界以非物质文化遗产展览和活动为特色，每天吸引几千人参观。邢同和发现，很多人特意来看这座历史建筑。

昨天：上海大世界　　**今天：**上海大世界　　**地址：**西藏南路 1 号

建造时间： 1917 年　　**设计师：** 周惠南

参观指南：早 8 点 - 晚 5 点（每周一二四五），早 8 点 - 晚 5 点，晚 6-9 点（每周六日），门票 60 元。建议登上天桥欣赏露天表演。

"Shanghai Dashijie is a condensed world," says architect Xing Tonghe, who headed the restoration of the century-old entertainment center that stands at the corner of Xizang Road and Yan'an Road M., close to the People's Square.

Dashijie, or the Great World amusement center, was opened in 1917 by Chinese businessman Huang Chujiu, who amassed a fortune from marketing a brain tonic. He founded the Great World after retreating from an earlier investment, the New World Amusement Center on Nanjing Road.

"Huang always knew how to cater to the Chinese people," says Zhang Yaojun, a researcher from Shanghai Archives Bureau.

"He faked a foreign doctor's name, Dr T. C. Yale, on the bottle of his brain tonic as Chinese tended to trust foreign inventions and products. And the Great World offered the best-value entertainment in town for Chinese who could spend a whole day in the palace at an admission price cheaper than a movie ticket. Favorite Chinese entertainment forms, such as acrobatics and traditional operas in different dialects, were its main fare."

The amusement center was a large building with a tower over the principal entrance, according to a story on July 21, 1917, in the North China Herald. A promenade completely surrounded a large open space for all kinds of amusement. There were many rooms set aside for diners, stalls, side shows and billiards.

"One of the great features is the provision for open-air amusement very much after the style of those places in Europe generally dignified with the title of 'Luna Park.' Here there will be roundabouts, where the merry celestial will experience all the thrills of a ride on a wooden horse, apparatus for aerial flights on the somewhat safe steel wire, a shooting gallery, panoramas reminiscent of the country fairs at home, cinematographic displays, a menagerie and a great wheel," the report stated.

The amusement center was reviewed as "a novel feature for China" though it was not so large, and its proportions were modestly small. Brilliantly lit with electricity both inside and out, the center formed an illuminated landmark at night.

The center offered all-day entertainment. The acrobatics were in the open-air theater while a variety of Chinese operas were staged in up to 10 small theaters. It attracted an average of 15,000 visitors daily, and the number amounted to 25,000 during holidays, mostly middle- or lower-class Chinese men.

The Great World did good business, but Huang's financial businesses were not running well. He was bankrupt after opening the Shanghai Day and Night Bank and was forced to sell all his enterprises to liquidate his debts. He died of a serious illness in 1931 at the age of 59.

Huang Jinrong, Shanghai's mob boss, took over the Great World the same year. He expanded the site, renovated the building and turned it into a comprehensive en-

tertainment venue featuring dining, stage shows and shopping malls.

The Great World was taken over by the Shanghai government in 1954 and renamed several times from the People's Playground, Dong Fang Hong Theater to Shanghai Youth Palace.

It reopened once again in 1987 as a popular entertainment venue with its old name, the Great World. In 1992 the center started the Great World Jinisi Records, which sounded similar to the Chinese translation for Guinness World Records, and invited people from all over the country to break world records.

However, as new entertainment venues such as theme parks and aquariums mushroomed and young people weren't interested in Chinese operas and folk art shows, the Great World was on the skids. In 2004 when many public places were closed down temporarily due to the outbreak

of the SARS epidemic, the Great World shut its doors as well for a decade-long restoration headed by Xing.

Many Shanghainese had fond memories of the Great World, especially 80-year-old architect Xing, who spent many weekends here when he was a primary school student in the 1940s.

"My grandfather, a senior employee of Li Dinghe Brush Pen Shop near Dashijie, often took me there to spend a weekend afternoon. While he was admiring a Chinese opera, I would explore this multi-level emporium of entertainment by myself," Xing recalls. "I climbed the balustrade staircases up and down, laughed in front of the distorted mirrors and watched performances in different theaters. Dashijie was so big in my mind when I was a boy."

But the Great World entertainment center, dilapidated and full of scars, turned out to be not so big when the architect began surveying it before restoration in 2004. "The railings were cracked, the reinforcement bars exposed and the floors warped, while the distorted mirrors were very dusty," recalls Xing. "After serving some 10,000 people every day for over 80 years, the Great World looked like an aging warehouse. I was so sad," he says.

Before the restoration kicked off, there was fierce discussion over its future role, whether to turn it into a trendy cartoon park or maintain its traditional essence. The architect believes it is a condensed world of Shanghai's entertainment and cultural life.

"The Great World was built for ordinary people since its very first day. It had also been a platform for young emerging artists who needed a place to perform and practice, some of whom grew to be stars afterward. It could be reverted to a modern cartoon park, but the original flavor would be lost forever. Cartoon city could be built anywhere," he says.

The veteran architect and his team restored the building according to the look during its prime time in the 1930s. They consolidated the endangered structure, preserved the original mosaics the promenade and the balconies, and rebuilt demolished pavilions on the top, according to historical photos and original drawings. While preserving historical details, they also took modern function into consideration.

At its centennial memorial, the Great World reopened on March 31, 2017 to exhibit intangible cultural heritage and attracted thousands of people every day. Many of them, Xing says, came to see the famous architecture.

Yesterday: Shanghai Great World   **Today:** Shanghai Great World
Address: 1 Xizang Rd S.   **Date of construction:** 1917   **Architect:** Zhou Huinan
Opening hours: 8am-5pm, Mondays-Tuesdays, Thursdays-Fridays; 8am-5pm, 6-9pm, Saturdays-Sundays   **Admission:** 60 yuan
Tips: Climb onto the promenade to admire an open-air performance just like architect Xing Tonghe did when he was a boy in the 1940s.

蓝色琉璃的光芒

Harmony of East and West

1931年，耗资百万的八仙桥基督教青年会大楼落成，被誉为"将中国建筑融入高层结构的首次成功尝试"。这座中西合璧的大楼至今保存完好，中式屋檐上的蓝色琉璃瓦在上海市中心熠熠生辉。

青年会大楼开幕前，英文《大陆报》称这是"一次中西风格的和谐融合"，是"远东地区这种类型建筑里最现代的之一"。

中国古代建筑多为木结构，很少建有高层建筑。20世纪20年代末，民国政府提倡复兴中国传统文化，一批中国建筑师尝试采用西方先进的建筑科学技术，将中国传统建筑元素融入高层建筑。这种中西合璧的风格被学界称为"中国传统复兴风格"，青年会大楼是一个经典案例。

"大楼底部用预制的人造条石筑成，上部则以红色泰山面砖覆盖，顶层饰有柱、梁、格子窗和中国庙宇屋顶常用的琉璃瓦。"《大陆报》写道。

这座9层大楼是上海基督教青年会的新厦。基督教青年会1844年由英国人乔治.威廉姆斯（George Williams）创办，是一家遍及世界各地的跨宗教社会组织。1900年，一批华人精英创办了上海基督教青年会，发起者包括外交官颜惠庆和宋氏三姐妹的父亲——宋耀如。

虽然基督教青年会带有宗教背景，但上海基督教青年会侧重开展针对青年人的各种文体活动。青年会在新大楼举办讲座、音乐会、戏剧表演、艺术展览和体育运动，包括鲁迅在内的很多名人都曾来这里举办讲座。

抗战期间，爱国活动在上海屡屡被禁，吴耀宗、胡愈之、王任叔和雷洁琼等知识分子在青年会九楼举办"星二聚餐会"，以聚会的形式举办宣传团结抗日的论坛讲座。美国著名记者安娜·路易斯特朗(Anna Louise Strong)和埃德加·斯诺（Edgar Snow）都曾经发表演讲。正是因为"星二聚餐会"成员的推介赞助，斯诺介绍中国革命的成名作——《西行漫记》（*Red Stars over China*）出版了影响深远的中文版。

1949年后，大楼改为国营淮海饭店，1978年大楼产权又回归青年会。上海基督教青年会于1984年恢复昔日功能，提供教育文化等多种社会服务，如今在大楼顶层办公，将其余楼层租给锦江集团旗下的精品酒店——都城青年会酒店。

这座中西合璧的大楼保存完好。走进石雕拱门好像走进一个小巧的中式宫殿。大厅以藻井和护墙板装饰，明亮多彩，宛若1931年一名外国记者在大楼开幕前探班时所见的样子。

"这种结合既体现了令人愉悦的西方建筑比例，而在重要特征的细节上又严格遵循品位高雅的中国风格。可以说，大楼将中国建筑设计为现代高层，就算不是这种成功尝试的孤例，也是案例之一。而此次尝试也证明这种建筑不仅是一种考古学研究，而且是一种可变为现实的建筑风格。"《大陆报》记者写道。

这件成功的作品由三位曾在美国学习的中国建筑师——范文照、赵深和李锦沛共同设计完成。其中，范文照和赵深毕业于美国宾夕法尼亚大学，曾一起设计了南京大戏院（今上海音乐厅）。

而李锦沛的人生轨迹与他们不同。李锦沛于1900年出生于纽约，曾在麻省理工学院和哥伦比亚大学学习，后来获得纽约州立大学学位。他毕业后曾在美国建筑师茂飞的事务所工作，而茂飞后来在中国设计了大量中国传统复兴建筑，包括南京灵谷寺阵亡将士纪念塔和燕京大学（今北京大学）。1923年，李锦沛到上海担任基督教青年会建筑处的副建筑师。这位美国土生土长的建筑师却在上海设计了不少中国风浓郁的作品。西藏路青年会大楼竣工一年后，他在圆明园路又设计了8层高的基督教女子青年会大楼，也是一个融汇中西建筑风格的作品。女青会大楼装饰有须弥座和回纹，藻井雅致的大堂与西藏路青年会大楼有异曲同工之妙。

研究近代中国建筑师的同济大学钱锋教授提到，李锦沛在中国设计了一系列基督教青年会建筑。1929中山陵设计师吕彦直因

病去世后，李锦沛接过重任将未竟的设计任务完成。而李锦沛、吕彦直、范文照和赵深都曾经在纽约茂飞事务所工作过。传统中国建筑行业只有工匠和手工艺人的概念，李锦沛这一批华人建筑师被称为近代中国第一代建筑师。他们中的很多人20世纪初曾赴欧美留学，回国后将所学技艺用于建设正在向现代化转型的中国。

不过，李锦沛、范文照、赵深和其他很多同时代的中国建筑师，后来都转向更为现代的风格。李锦沛位于南京西路的作品华业大楼就是一座简洁的西班牙式公寓楼。

2015年，美国华侨博物馆举办了介绍李锦沛的特别展览。这位来自纽约唐人街的设计师在1945年因战乱离开上海回到纽约。他曾经把在美国学到的西方建筑技术引入中国，后来又把中

国古典复兴风格从中国带到纽约唐人街，用在很多作品上。"李锦沛是第一个为唐人街美国华人做设计的美国华裔建筑师。"纽约特展如此评价他。

这个展览题为"中国风格，再发现李锦沛建筑1923—1968"，既展示了这位年轻的建筑师与同事在中山陵前的合影，也有他为纽约唐人街项目设计的草图。其中很多项目都有着中式大屋顶和彩绘装饰，与那座万里之外的蓝色琉璃瓦的酒店，非常神似。

昨天：八仙桥基督教青年会大楼　**今天：**锦江都城青年会精品酒店＼上海基督教青年会
建造时间：1931年　**建筑师：**范文照、赵深、李锦沛　**建筑风格：**中国古典复兴风格
参观指南：酒店大堂有关于建筑历史的小型展览。建议也参观建筑师李锦沛的另一件作品——圆明园路133号女青年会大楼。两座建筑的保护状况都不错，室内有丰富的历史细节。

The nine-story Chinese YMCA Building was regarded "a first successful attempt in which Chinese architecture is incorporated in high structure" in 1931. Today, it still stands tall, as a boutique hotel, with blue glazed tiles glistening in the heart of Shanghai.

When "the new $1 million Chinese YMCA building" was nearing completion in the August of 1931, *the China Press* called it "a harmonious composite of East and West" and "one of the most modern of its type in the Far East".

"With the exception of pagodas, old Chinese buildings, most of which were made of wood, rarely have more than two floors," says Tongji University professor Qian Zonghao.

"But starting from the late 1920s, the Chinese government began to call upon Chinese to revive traditional culture, and Chinese architects applied their Western knowledge to design taller buildings with traditional elements. We call it 'Chinese Renaissance Style'."

The Chinese YMCA Building is a classic example of this style. According to the China Press, the lower level forming the base of the building is constructed of pre-cast artificial stone ashlars, the upper part being faced with Taishan red face bricks and the top story treated with the effect of pillars and beams, latticed window work and the usual Chinese roof of glazed temple tiles.

The building was the new edifice for the Chinese Young Men's Christian Association. The history of the YMCA dates back to 1844 when Englishman George Williams founded the international organization.

The YMCA in Shanghai was established in 1900 by a galaxy of Chinese elites and sponsors, including diplomat Yan Huiqing and prominent businessman Charlie Soong, father of the famous Soong sisters.

Although an institution with a religious background, the YMCA was not engaged in preaching religion but carried out various activities focusing on young people. The new building hosted numerous lectures, concerts, drama performances, art exhibitions and sports activities. Many celebrities including famous Chinese writer Lu Xun hosted talks here.

During China's War of Resistance Against Japanese Aggression (1931-1945), Chinese intellectuals initiated "Tuesday dining

party" in the restaurant hall of the building.

These functions, in the name of dining parties, were informal forums that united people to engage in the underground fight against Japanese aggression.

Among the speakers were famous US correspondents Anna Louise Strong and Edgar Snow. The Chinese translation of Snow's "Red Star over China" , an influential account of the Chinese revolution, was published with the support of "Tuesday dining party" members.

After 1949, the building became the state-owned Huaihai Hotel, but ownership was handed back to the YMCA in 1978, which still has an office inside but rents out most of the building to the Jinjiang Metropolo Hotel Classiq, a boutique hotel.

"The Chinese YMCA resumed its function in 1984 and is more of an organization today for social services," says Ma Zhaozhen, director of YMCA Shanghai.

"We provide services ranging from education to culture, sports and day care in more than 10 community centers across the city. Every year, 3 million people-from children to youngsters and the elderly-benefit from YMCA services and activities, more or less changing their lives."

Today, most of historical features are well-preserved. Walking through the stone-arched gate of the hotel is like entering a small Chinese palace. Most rooms are decorated in Chinese style, with tastefully decorated beamed ceilings that are set off with wood-panelled walls, retaining it's original style.

"The composition shows a pleasant nicety of proportion of the Western order and yet every detail in the essential features and ornamentations is worked out strictly in accordance with the Chinese style of good taste," a journalist for the China Press wrote in 1931.

"It can be safely said that this building is one of the first, if not the only, successful attempts to create a modern tall building in the Chinese architectural style and to prove that this style of architecture has become not only a study of archaeology but a living style of architecture."

The building was designed by a trio of US-trained Chinese architects--Poy Gum Lee, Fan Wenzhao and Zhao Shen--says Tongji University Associate Professor Qian Feng.

Fan and Zhao, both graduates from the University of Pennsylvania, also co-designed Nanking Theatre which is today the Shang-

hai Concert Hall.

"Unlike them, Lee was born in New York in 1900 and studied architecture in Price College, MIT, Columbia University and won a diploma from New York State University," says Qian.

Before coming to Shanghai as a YMCA architect in 1923, Lee had worked in the New York office of American architect Henry Murphy, a foremost architectural proponent of the incorporation of Chinese architectural elements into modern construction.

"Familiar with Murphy's manners and treatments, Lee later used the new style extensively in his China projects including the Chinese YMCA Building," Qian says.

Although born and educated in the US, Lee's architectural educa-

tion in several renowned American universities coincided with the emergence of this modern Chinese style.

A year after the Chinese YMCA Building was completed in 1931, he adapted Chinese architectural elements and motifs to the eight-story bureau for the National Committee of the Young Women's Christian Association of China which still stands on Yuanmingyuan Road near the Bund.

Qian notes that besides a rainbow of YMCA buildings across China, this prolific architect also took over the work for constructing Dr Sun Yat-sen's Mausoleum in Nanjing, capital of Jiangsu Province, a typical example of Chinese renaissance style, after chief architect Lu Yanzhi suddenly died of disease in 1929.

Lee, Lu, Fan and Zhao had all worked in Murphy's office in New York. They were regarded as modern China's first generation of architects, since building in China traditionally adhered to regional traditions, and architecture did not then exist as a profession.

During the first decades of the 20th century, many young Chinese were sent abroad to America and Europe and later brought back skills to help modernize China.

"Like Fan, Zhao and other Chinese architects, Lee later turned to more modern style, and a noteworthy work is the Cosmopolitan Apartments, a simply cut Spanish-style building with Art Deco details hidden in a lane of Nanjing Road W.," Qian says.

However, Lee's successful career in China was cut short by the Japanese invasion, and in 1945 he repatriated to design projects in New York's Chinatown where he was born and grew up. This time he brought architectural modernism from China to New York and became the first Chinese-American architect to design for clients in Chinatown.

To commemorate Lee who bridged two cultures, the Museum of Chinese in America in New York hosted an exhibition titled "Chinese Style, Rediscovering the Architecture of Poy Gum Lee

1923-1968" in 2015.

The exhibition displayed a photo of Lee with a colleague fronting the grand Mausoleum of Dr. Sun Yat-sen and his drawings of Chinatown projects. Most of these buildings featured big roofs and colorful paintings that resembled the blue-tiled hotel in the heart of Shanghai.

Yesterday: YMCA Building   **Today:** Jinjiang Metropolo Hotel Classiq
Address: 123 Xizang Rd S.   **Date of construction:** 1931
Architects: Fan Wenzhao, Zhao Shen and Poy Gum Lee
Architectural style: Chinese Renaissance   **Tips:** The hotel has a mini exhibition featuring archive photos of the building's rich history near the lobby. Poy Gum Lee's other works in the Chinese renaissance style and the YWCA building at 133 Yuanmingyuan Road are also worth a visit. Both buildings are well-preserved and feature amazingly beautiful interiors.

1931年范文照关于中国建筑的演讲
Fan Wenzhao's Speech on Chinese Architecture

昨晚在西侨青年会乔治·费奇（George A. Fitch）主持的会议上，知名中国建筑师范文照先生做了一个非常有趣的演讲。这位曾经创办中国建筑学会的获奖建筑师简要追溯了中国建筑的历史，将其与西方建筑历史做了比较，并概括了中国建筑的原则与特征。

在比较中西建筑时，范先生告诉听众，除了大约是元朝忽必烈皇帝统治的时期，中国人从来都不是崇尚武力的，因此不像其他国家一样会建造凯旋门和战争纪念碑。而且，他继续谈道，中国人总体而言也并不热衷于宗教，所以中国没有伟大的庙宇，不像欧洲有那么多壮观美丽的天主教堂和基督教堂。

他又指出，在传统中国社会，艺术家和建筑师就职业而言是学者和诗人。他们从事艺术研究，以此作为个人实践与表达的方式。而在欧洲中世纪时期，艺术家的地位更加显要。在意大利，他们会得到如美第奇家族这样的金主授予的荣誉。

范先生说中国建筑大约始于晋朝，但具体时间还缺乏文献考证。公元前214年，最著名的中国建筑工程——有1149英里长的长城由秦始皇建成。范先生指出，一直以来，中国房屋特别是宫殿都是在北、西、东三面筑墙，只留南面开放，设为前门。建筑的每个元素都有结构上的意义和装饰作用，也非常实用。与南方建筑相比，北方建筑的曲线被有效处理为精妙的效果。而在南方，曲线发展为一种太大的形式。用于装饰作用的人物和动物形象有宗教意义，他们也被当作驱除邪魔的保护神。

演讲者极为重视中国建筑的复兴，他指出了一个事实，即过去中国的艺术家一直在效率与美之间的挣扎中比较东西方。但在最近，他解释道，出现了一小批新的艺术家，追求将东西方艺术的优点相结合。这些现代中国的建筑师开始用微薄的努力，对抗那些对西方建筑的丑陋模仿。他们努力将现代便利与舒适引入保留着中国传统美的建筑中。

Fan Wenzhao, one of the three architects who designed the Chinese YMCA Building in Shanghai traced the history of Chinese architecture and talked about China's architectural renaissance in an address in 1931 when the building was nearing completion:

"Briefly tracing the history, principles and peculiar features of Chinese architecture and comparing it with the West and outlining the principles and peculiar features of the former, the noted Chinese architect, winner of several prizes and founder of the Chinese Architects' Society delivered his speech with lantern illustrations at the meeting presided over by (American missionary and YMCA official) George A. Fitch at the Foreign YMCA."

In comparing Chinese architecture with that of the West, Mr Robert Fan told his audience that the Chinese were never military people except perhaps during the reign of the Yuan Emperor Kublai Kahn and consequently there are no triumphal arches and war memorials like those in other countries.

He also said the Chinese people as a whole had little religious zeal, and therefore China has no great temples like the great and magnificent cathedrals and churches in Europe.

In this respect, he further pointed to the fact that Chinese artists and architects in the old order were by profession scholars and poets who took up the study of art as means for personal experiment and expression whereas in Europe during the Middle Ages artists were looked upon as superior and men of great importance and were given high honor by great patrons of art like the Medici in Italy.

About 214 B. C., the most famous of Chinese building undertakings, the Great Wall, was built by Emperor Qin Shi Huang. It is 1,140 miles (1,824 kilometers) long. The Tang Dynasty (AD 618-907) and Song Dynasty (960-1279) figured as the flourishing stages of Chinese architecture with emperors as great patrons.

Mr Fan pointed out that almost always Chinese houses and especially palaces have their walls on the north, west and east only leaving the south for their front doors.

In grouping houses a strict balance is always desired. In point of sincerity there is no false idea of construction--each element in the structure has its structural value and decoration and is a matter of inspired utility.

The subtlety of the curve is more effectively preserved in the Northern style than in the south where it has been developed into too fantastic a form. Human and animal figures used for decorative purposes have religious values in as much as they are regarded as protecting spirits against evil demons.

The speaker attached paramount importance to the Chinese architectural renaissance: He pointed to the fact that in the past in China, artists had been comparing the East with the West in the light of a struggle between efficiency and beauty.

Recently, however, he explained there has appeared on the stage a new but small group of men who seek to bring about a synthesis of the best in both.

They are the architects in modern China who are beginning to make their feeble efforts felt in fighting against the ugly imitations of Western

Architecture and endeavoring to introduce the modern conveniences and comforts into buildings that can still retain the old beauty of China.

摘自 1931 年 5 月 13 日《大陆报》

Excerpt from *the China Press*, on May 13, 1931

西洋剧场的中国设计

A Western Theater designed by Chinese Architects

上海音乐厅是一座欧洲古典主义建筑，却是由两位中国建筑师——范文照和赵深设计的。

1930年3月26日这座建筑开业时是一家名为"南京大戏院"的现代影院，位于今天的延安东路。开业数月前的1929年10月，英文《大陆报》就刊登预热报道，隆重介绍这家造价50万元的

新电影院体现了"最新式的建筑特点"。

"无论从艺术装饰的角度，还是从让观众舒适的方面，整座建筑都展示了很多新颖之处。工程耗资超过50万元，这还不包括基地的费用。影院最新潮的地方就是安装了空调系统，观众厅的空气经过滤除尘后在室内循

环，温度常年保持华氏70-80度（约摄氏21-27度），空气相对湿度也通过科学的控制与调节。这些设备与纽约罗可西剧场和帕拉蒙剧场所使用的设备非常相似。"1929年的报道写道。

南京大戏院由华商联怡公司Shanghai Amusement Co. Ltd投资兴建。影院除了空调，还装有用于播放有声电影和维他风（用唱片伴音的早期有声电影技术）的西洋电子声音播放器。

这座影院建筑设计为改良的文艺复兴风格。外立面的拱廊由彩色灰泥和人造石建造而成。北厅设计有古罗马柱式和大理石楼梯，剧场内有巨大的穹顶和雕刻精美的护墙。南京大戏院开业后，除了放映电影，还上演杂技、马戏、戏剧和地方戏等多种节目。

"拱廊顶部是一块雕有合适主题的浅浮雕镶板。两侧立面饰有泰山面砖和人造石。观众厅里设计有爱奥尼克壁柱和拱券，由彩色打褶布帘装饰。天花板将由多彩的浅浮雕装点。照明是非直射的光线。"报道写道。

有趣的是，南京大戏院的两位中国建筑师在完成这件古典之作以后，设计风格都发生了显著变化。他们先创作了一些将中国元素融入现代建筑的中国古典复兴建筑，但后来都转向更纯粹简洁的现代主义建筑。这两位曾留学美国宾夕法尼亚大学的建筑师也都成为中国第一代建筑师里的代表人物。

赵深离开范文照事务所独立执业后，参与创办著名的华盖事务所，1932年担任中国建筑师学会会长。华盖事务所是老上海最著名的两家大型中国建筑事务所之一，另一家是设计南京路大新公司（现一百商业中心老楼）的基泰工程司。

研究近代中国建筑师的同济大学钱峰教授评价，赵深的设计风格很像他本人的个性——"稳重朴实、美观大方、注重功能与节约建造成本"。

"设计南京大戏院的两位中国建筑师——赵深与范文照，早年曾分别参加南京中山陵设计竞赛。他们都获得了奖项，但中山陵最终由另一位中国建筑师——

吕彦直设计，"钱锋说。

作为近代中国的第一代建筑师，赵深和范文照设计了不少中西合璧的"中国古典复兴建筑"，如带中式屋顶的八仙桥基督教青年会大楼。

"但他们都抛弃了这种风格。赵深所在的华盖事务所曾发起'摒弃大屋顶运动'。而范文照转向现代风格后，呼吁纠正为建筑添加传统大屋顶的错误。范文照的风格转变是受到两位新同事——美籍瑞典裔建筑师林朋（Carl Lindbom）和中国建筑师伍子昂的影响。1935年他曾代表中国参加在英国伦敦举办的第14届国际住房与城镇规划大会和在意大利罗马举办的第13届国际建筑大会，回国后更坚定地拥抱了'功能决定形式'的国际式风格。范文照1941年设计的美琪大戏院就非常现代。" 钱锋说。

除了赵深和范文照，同时代的一批中国建筑师都纷纷转变了设计风格，南京大戏院也因此成为上海现存不多的由中国建筑师设计的西方古典主义建筑。

在20世纪30年代，这座外观古典、设施现代的影院为观众提供极致舒适的观影环境，并总是上映最新的影片。根据范文照孙女Maureen Fan的回忆，祖父外国朋友多，在家中多讲英语，起居室的椅子是德国现代主义大师——密斯·凡·德罗的作品。

"周末，祖父会带我的父亲、叔叔和两个姑姑去这家他设计的戏院免费看马克兄弟（Mark Brothers，美国早期喜剧演员）或约翰·韦斯穆勒（John Weissmuller 1904-1984，美国游泳运动员、演员）的《人猿泰山》。" Maureen Fan 在2009年《华盛顿邮报》的一篇文章中写道。

1950年，南京大戏院改名北京电影院，1959年因其出色的剧场音效，变为上海音乐厅，成为向大众介绍古典音乐的文化场所，常年上演交响乐、室内乐、独奏、爵士乐、演唱会等。很多世界一流乐团和音乐家，如费城室内乐团、巴伐利亚广播交响乐团、小提琴家祖克曼和钢琴家波利尼都曾在历史悠久的剧场登台献演。

2002年，为配合市政建设，这座历史建筑在被提升后向西南方向平移66米，随后进行修缮和扩建，2004年坐落于延安中路绿地内的音乐厅重新开幕。

负责平移和修缮音乐厅的建筑师章明回忆，工程不仅对建筑立面进行清除污渍和修补缺损，还将原先仅100平方米、进深8米多的舞台扩充到298平方米，进深增加到14米，并新增了沉降乐池。此外，观众座位和休息空间也大为增设与改进。

如今，这座音乐厅很多地方，看上去仍然如同1929年英文报纸描述的模样：

"大堂的大理石楼梯通向一个位于夹层的宽敞走廊。在这里，大理石柱坐落于石基座上。天花板装饰华丽。在建筑东侧，会有一个连接大堂圆厅的休息厅。在夹层走廊上，透过大理石柱廊可以看到大堂。"

章明提到，天花板、圆厅和墙都根据历史原貌修复，室内部分则精心选用了金色、灰蓝色和

米色装饰，以求与这座古典建筑的风格相配。

上海音乐厅总经理方靓认为，这座建筑不仅是一座音乐厅，也不只是一座20世纪30年代的建筑，它还有一个漂移的故事。

"我在这里工作十几年，音乐厅每周有演出，观众进来都会不停地拍照，看墙上音乐厅历史的介绍。要了解上海，上海音乐厅就是一个切入口。我们有责任把故事讲好并保护好建筑，让更多人在这里不仅能欣赏高雅艺术，同时走上80多年的台阶，在抚摸长廊扶手的时候，能感受到这座城市的温度和历史厚度。"她说。

昨天：南京大戏院　**今天：**上海音乐厅　**建筑师：**范文照、赵深　**建筑风格：**西方古典风格

参观指南：建议演出前提前一小时抵达，欣赏建筑，并在西厅咖啡馆喝一杯。西厅为平移工程时扩建的部分。

演出信息请查音乐厅官网：www.shanghaicncerthall.org.

Shanghai Concert Hall, a rare surviving example of a Western classic building designed by Chinese architects in old Shanghai, was the work of Fan Wenzhao and Zhao Shen, both graduates of the University of Pennsylvania.

"They are representative figures of China's first-generation modern architects," says Tongji University associate professor Qian Feng.

The three-story concert hall was built in 1930 as the Nanking Theatre, a cinema, on today's Yan'an Road E..

The China Press called it "Shanghai's New $500,000 Cinema" and said "its structure embodies most up-to-date features in construction" in October 1929, ahead of its official opening on March 26, 1930.

"The entire structure, which reveals many novel features both in point of artistic decoration and those designed with a view to the comfort of patrons, is being erected at a cost of more than $500,000 exclusive of the cost of the site," the report said. "The most novel of these features will be the installation of an air-conditioning plant, by which air washed free of all dust will be circulated through the auditorium, at a temperature of from 70 to 80 degrees Fahrenheit throughout the years and its relative humidity will be scientifically controlled and adjusted. This apparatus is quite similar to those installed in the Roxy and the Paramount theaters in New York."

Invested by Chinese-owned Shanghai Amusement Co. Ltd., the cinema was equipped with Western electric sound projectors for Movietone and Vitaphone films.

The building's design is modified renaissance. The facade features arcades built with colored stucco and artificial stones. The structure had ancient Romanesque pillars and marble stairs in the north hall, a huge dome inside, as well as parapets with delicate carvings. During the early years of the theater, movies, acrobatics, circuses, dramas and traditional Chinese operas were put on here.

The China Press report said: "The arcades are surmounted by a sculpture panel with a suitable theme in relief. The side elevations are in Taishan face brick and artificial stone. The walls of the auditorium will be treated with a series of Ionic pilasters and arches decorated with rich draperies. The ceiling will be in rich, low relief.

The lighting will be indirect."

Professor Qian notes that both the architects changed to Chinese renaissance and then to an utterly modern style after creating the Western classic building.

Zhao returned to China in 1927 to join Fan's firm. He later co-founded the Allied Architects, one of the two leading Chinese design firms in modern China--the other being Tianjin-based Messrs Kwan, Chu and Yang, designers of the Shanghai No. 1 Shopping Center on Nanjing Road E..

"Zhao's style is pretty much like his personality--simple, sedate and elegant, stressing functionality and economic construction. It's interesting that both Zhao and Fan took part in the competition for designing Dr. Sun Yat-sen's mausoleum in Nanjing, Jiangsu Province. They both won prizes, but the mausoleum was finally designed by another Chinese architect named Lu Yanzhi," Qian says.

As modern China's first-generation architects, Zhao and Fan designed some "Chinese renaissance" buildings which incorporated both Chinese and Western elements, such as the YMCA building with its upturned eaves and large plate-glass windows.

"And they both gave up the style," Qian says. "Zhao's firm launched a campaign to abandon big Chinese roofs while Fan seemed even more radical. He called on correction of this 'big roof mistakes' , especially after European architect Carl Lindbom and Chinese architect Wu Zi'ang joined his firm."

"After his 1935 tour around Europe representing China at the 14th International Housing and Town Planning Congress in London and the 13th International Architectural Congress in Rome, he fully embraced 'international-style' architecture that valued the concept of 'form follows function.' His 1941 work, the Majestic Theater on Nanjing Road W., was a very modern piece," she says.

According to Fan's granddaughter Maureen Fan, he spoke English at home, counted foreigners among his friends and had chairs designed by German modern architect Mies van der Rohe in his living room.

"On weekends, he took my father, uncle and two aunts to see the Marx Brothers or Johnny Weissmuller's 'Tarzan' free of charge in theaters he designed," Fan wrote in an article published in the Washington Post in 2009.

In 1950, the theater was re-

named the Beijing Cinema and in 1959 embraced a new name and new role as the Shanghai Concert Hall. Since then, it has been a popular cultural venue specialized in presenting classical music, jazz and traditional Chinese music.

The theater co-hosted the city's major cultural events including the Shanghai Spring International Music Festival and the Shanghai International Arts Festival. World-class troupes and artists, including the Juilliard String Quartet, the Bavarian Radio Symphony Orchestra, violinist Pinchas Zukerman and pianist Maurizio Pollini, have performed there.

To make way for urban construction, the building was relo-cated in 2002. It was lifted and moved 66 meters southeast to sit in the Yan'an Road M. greenland before reopening in 2004. Some expansion and renovation was carried out during the project.

"At the same time as cleaning and restoring the original facade, we enlarged the stage from 100 to 298 square meters and added a sunken orchestra pit. Seats and modern facilities for artists and audiences to relax were also added during the project," architect Zhang Ming wrote in an article.

Today, the concert hall still looks like the building *China Press* described in 1929.

It said then: "The main lobby is characterized by a marble staircase giving access to a spacious promenade on the mezzanine floor. Here marble columns will rest on a cain-stone base course. The ceiling will be richly ornamented. There will be a foyer on the east side of the building connecting with the main lobby beneath a beautiful rotunda. The main lobby may be viewed from the mezzanine floor promenade through a marble colonnade."

Zhang recalls the ceiling, the rotunda and walls have been repaired with original materials according to their old look. Elegant tones of gold, grayish blue and beige were selected for the interior decoration to suit the ambience of the building.

"The building is more than a concert hall. It's not only a 1930s building, but also has a story of floating and relocation. I've worked here for more than 10 years and often find our audiences enjoy taking photos of the building before or after a concert," says Fang Liang, the venue's general manager.

"This building is an entrance for people to get to know about Shanghai. We have a responsibility to preserve it well and share its stories. I hope visitors to Shanghai Concert Hall admire elegant music, meanwhile walk up the 80-year-old staircase, touch the handrails of the veranda and feel the warmth and depth of our city," she says.

Yesterday: Nanking Theatre   **Today:** Shanghai Concert Hall   **Address:** 523 Yan'an Rd E.
Date of construction: 1930   **Architects:** Fan Wenzhao and Zhao Shen
Architectural style: Western classic style
Tips: The concert hall is open only to its audiences. I'd suggest you visit the building one hour ahead of a performance to admire it and have a drink at the cafe in the western hall, which was built during the relocation project. For ticketing information, visit www. shanghaiconcerthall.org.

真正的建筑无需花边
Modern architecture "needs no frills"

2月9日，卡尔·林朋先生（Mr. Carl Lindbom）在范文照事务所为他举办的招待会上强调，"现代建筑无需花边"。他已经加入事务所成为一名合伙人。招待会参加者众多，林朋先生用美丽的模型诠释新的"国际式建筑"，他是这种建筑的倡导者。

他说，这种结合了美与简洁的新建筑风格，已经取代昂贵的复古风格，在全世界快速流行。在19世纪和20世纪，建筑大多模仿早先时期的风格，在表面堆砌装饰，这些装饰与内在并没有关系。

另一方面来说，他并不认同"现代建筑"只是关于垂直和水平线条、闪亮金属和多种装饰的尝试，这些都是"太肤浅的事情"。

国际式风格展示了一种由建筑本质而决定的建筑类型。举个例子，林朋设计的talkie工作室几乎是半圆形的。采用这种形状所有的支撑和梁都不需要，整幢建筑由混凝土建成，看上去井然有序，外部没有花哨的装饰。

另一个案例是一座公寓楼，由钢、混凝土和玻璃建成，每套公寓就好像主楼伸出的一个架子一样。无论什么朝向，光线、空气和阳光能从三面进入公寓。同样没有花哨的装饰，并不需要。

林朋先生曾在哥本哈根皇家艺术学院和瑞典学习建筑，师从现代建筑大师格罗皮乌斯（Walter Gropious）、柯布西耶（Le Corbusier）、吕夏（Andrew Lurcat）和赖特（F. L. Wright）。

他的作品包括圣莫尼卡科特先生（H.C. Cotton）的伯恩海默花园（Bernheimer Gardens）。他也曾担任美国加州克莱蒙特市建筑师和耗资200万美金的贝纳尔伯特寺（Bennar Berit Temple）的主设计师，这座建筑被视为美国意大利风格建筑的最佳范例。此外他的作品还有好莱坞的诺曼底村，很多明星安居于此。他刚完成内华达州拉斯维加斯第一家国际式风格的酒店方案，将有1000个房间，包括游泳池、舞厅、剧场、赌场，还有延伸的庭院与马厩。

　　"新式建筑很快就会出现。" 林朋推断道。他会让金融家们认识到这种国际风格所具有造价经济和简洁美观等优点，而这些正是他们所需要的，甚至超出了预期。

That modern architecture needs no frills was stressed by Mr. Carl Lindbom on Feb. 9 at a reception given in his honor at the office of Mr. Robert Fan, with whom Mr. Lindbom has gone into partnership. The reception was well attended, and Mr. Lindbom, with the aid of beautifully made models explained the new "International Style of Architecture" of which is he an exponent.

The style, Mr. Lindbom said, is rapidly becoming universal as combining beauty with simplicity, it is sweeping away older, more expensive styles. Early in the 19th and 20th centuries, he went on architecture aimed at the imitation of former periods by the use of "motives and ornaments placed on the body of the buildings without having an inner connection with them."

On the other hand, Mr. Lindbom does not approve of "Modern Architecture", which is "not architecture" but contains various experiments in vertical and horizontal expression, shiny metals and various ornaments, all of which are "largely superficial things".

The international style, however, demonstrates a type, which, not existing for its own sake, is determined by the particular nature of the building it is intended for. For instance, a "talkie" studio, designed by Mr. Lindbom, is roughly in the form of a semi-circle. By the use of this form of construction all supports and beams are done away with, the entire building being executed in concrete and sternly business-like has no external frills.

Another instance is an apartment building, constructed with steel, concrete and glass and so built that each apartment juts out somewhat like a shelf from the main building and being made of steel and concrete requires outward support. Light, air and sunshine thus are able to play upon all three sides of the apartment, no matter in which direction it faces. There are no external frills, none being necessary.

Mr. Lindbom, completing his schooling at the Royal University, Denmark studied architecture at the Royal Art Academies of Copenhagen and Sweden. At one time or another he has been a student of such modern architects as Walter Gropious, Le Corbusier, Lurcat and F. L. Wright.

Some of his works include the Bernheimer Gardens at Santa Monica on the estate of Mr H. C. Cotton; architect of San Clemente City, California, chief designer of the Bennar Berit Temple, which costing G$2,000,000, is exploited as the finest example of Italian architecture in America; the Normandy Village, Hollywood, home of many of the stars; and he has just completed the plans for the first hotel in the international style to be built at Las Vegas, Nevada, which will have a capacity of 1,000 rooms, and will include swimming pools, ballrooms, theater, casino and very extensive grounds and stables.

"New buildings will soon be undertaken," concluded Mr. Lindbom and he would draw to the attention of financiers the cheapness, simplicity, and beauty of the international style, which would be all they would require for their purposes and more.

摘自 1933 年 2 月 15 日 《北华捷报》

Excerpt from *the North-China Herald*, on February 15, 1933

大上海的摩天梦

A Skyscraper and An Architectural Adventure

2014年12月1日，上海国际饭店饼屋贴出一张彩纸通知，"因80周年店庆所有产品打八八折"。店外排起长龙抢购著名的蝴蝶酥等西点，老饭店用这种方式低调地庆祝落成80周年。

时光倒流到1934年12月1日国际饭店开业那一天，英文《大陆报》制作了一份特刊。这座中国第一座摩天大楼巨幅照片气势逼人，占据了几乎整个头版，其余20多个版面的报道从各角度详尽介绍这座位于上海市中心跑马场的大楼。这是一幢22层高、由四行储蓄会出资建造的大酒店，83.8米的高度堪称当时的亚洲最高楼。来自各国的建材设备供应商，如德国品牌西门子和国产的泰山面砖也纷纷刊登广告，骄傲地宣布这座堪称"上海历史上最有雄心的建筑冒险"的巨厦使用了自家产品。

其中有一篇报道透露，国际饭店能建这么高，与斯裔匈籍设计师邬达克（Laszlo Hudec）有关。如果不是这位东欧建筑师坚韧智慧的"斗争"，国际饭店就不会成为"远东第一高楼"。

邬达克在上海居住了将近30年，为这座城市留下近百个建筑作品，至今仍然熠熠生辉。

1914年邬达克毕业于布达佩斯的匈牙利皇家约瑟夫技术大学建筑系，但一战的爆发让美好前途成为泡影。年仅21岁的他应征入伍成为一名奥匈军队的炮兵，1916年不幸被俄军俘虏。1918年辗转逃跑至上海。

当时的上海是全世界少数几个不需要护照就能居住的城市，对邬达克来说是一个理想的避难地。他身无分文腿伤糟糕，口袋里只有一张用假冒俄国护照换来的安全通行证，唯一的资产就是作为建筑师的技能，而这在上海这座经济腾飞人口剧增的城市正好派上用场。功底扎实的邬达克从绘图员做起，迅速崭露头角，几年后就当上了美商克利洋行的

合伙人，1925年在外滩创办了自己的事务所。

1930年代是邬达克职业生涯的黄金时期，他在南京路沿线设计了三个标志性建筑——远东第一高楼国际饭店、远东第一影院大光明电影院和俗称"绿房子"的远东第一豪宅铜仁路吴同文住宅。其中国际饭店一项足以让他名垂青史。这座装饰艺术派风格的摩天楼，在中国保持建筑高度纪录近半个世纪，一直被视为上海的城市地标。

据《大陆报》报道，由于上海的土地由江海淤积的泥沙形成，大多数建筑都会发生沉降，所以监管租界工程的工部局在国际饭店之前从来没有批准过建造这么高的大楼。他们提出如沉降风险、火灾隐患等种种借口，而业主对这个高度也有顾虑。

邬达克竭尽所能，他先让储蓄会的业主相信，建造摩天楼不仅是可行的，而且在地价昂贵的上海是非常明智的决定。得到业主的认可后，他又与工部局展开周旋，承诺从技术上消除各类安全隐患，如增加消防喷头和使用高质量的钢材，最终让方案通过审批，国际饭店得以顺利建成。

"对于建筑师邬达克来说，这是一场长期而艰难的战斗，但最终他获得了胜利。看看这个漂亮的堪称技术典范的工程，谁不会为邬达克先生这开拓性的工作而喝彩？毫无疑问其他人会接着去建造一座座摩天楼，而他们肯定也会回头看看国际饭店这个最好的'鼻祖'。"报道写道。

邬达克用400根33米长的美国松木桩来夯实国际饭店的地基，又克服了大量技术困难，终于在土质松软的上海第一次建成了这么高的建筑。但是，开工前的这场"战斗"的难度，恐怕不亚于这复杂的地基工程。

国际饭店轮廓修长，顶部层层收进，楼顶好似去掉顶端的金字塔，雄心勃勃地指向天空。建筑的造型是纽约摩天楼惯用的手法，强调垂直线条。立面是防霜的深棕色耐火砖，塑造了国际饭店深色而高雅的经典形象。

四行储蓄会壮丽的营业大厅位于底层，柱子由黑色玻璃包裹，楼上休息厅的墙面镶板则是

乌木和黑色大理石制成，天花板呈现抛光的橙黄色。在这里客人们可以坐在愉悦的空调环境里，一边享用饮料，一边观看赛马比赛。

客房的家具均用进口核桃木和柚木制成。14层有举办庆典和私人派对用的沙龙和烧烤屋，通往一个巨大的景观露台。15到19层都是为贵宾客人设计的套房，储蓄会主任吴鼎昌的套房就在19层。这些套房所用家具陈设极尽奢华，足以满足到访上海的名流们的需要。

上海要建摩天大楼是受到美国影响。据同济大学刘刚教授介绍，1929年大萧条之前全世界都洋溢着乐观情绪，上海的经济很蓬勃，中国政治也暂时稳定，这一时期的财富积累和文化繁荣都是上海摩天的动力。从1920年代末到30年代初，上海有一种长高的氛围，一批著名高层建筑纷纷涌现，如峻岭公寓（现锦江饭店贵宾楼）、毕卡第公寓（现衡山宾馆）和沙逊大厦（现和平饭店），国际饭店也是其中之一。

邬达克设计国际饭店的灵感很可能缘于1929年大萧条前夕的美国之行。他是一个勤奋认真的人，从年轻时代起每到一地旅游都要记大量笔记和建筑素描，以积累设计素材。在美国访学期间，刘刚教授发现邬达克灵感的来源很可能是位于纽约暖炉大厦（American Radiator Building）的布赖恩特公园酒店（Bryant Park Hotel）。位于美国公共图书馆附近的这家酒店，无论是建筑风格还是褐色面砖饰面这样的细节特征，都与上海国际饭店有着亲切的姐妹关系，只是规模略小而已。

"邬达克所看到的1920年代纽约大都市景象是一种巨大的视觉冲击和心理冲击。他会感觉到一种力量、一种雄心，对未来有更多的一份自信。这份对于未来和城市的崇拜，对于普通人都是自然而然地发生的。"刘教授说。

但他认为，聪明的邬达克把纽约酒店的形体进行简化以适应国际饭店位于上海跑马场的基地，他对于纽约大都市风不是简单拷贝，而是捕捉此类建筑精华

后的一种文化移植。

《邬达克》传记作者、意大利建筑历史学者卢卡·彭切里尼（Luca Poncellini）认为国际饭店凝聚了邬达克的各种天赋，是其建筑师生涯的巅峰之作，而国际饭店的落成让上海渴望摩天大楼的梦想成为现实。

"在20世纪二三十年代之交（就像现在一样）摩天大楼是现代化的标志，象征了一个城市在国际舞台上的权势与成功。国际饭店楼顶飘扬的旗帜不仅标志着上海城市的最高点，同时表明这座城市开始以令人震惊的速度，去征服新的荣耀的高点……几十年来的梦想将由这座新近落成的壮美建筑来实现了。建筑大胆的高度及各处的配置显然反映了这个摩登时代……虽然与纽约、华盛顿和其他美国城市的摩天楼相比，国际饭店只有其几分之一的高度，但却是东半球——从伦敦到巴黎，建造过的最高的大楼，确实值得称道。"彭切里尼在书中写道。

1934年《大陆报》的报道也认为国际饭店是将上海由一座普通城市向"领先城市"提升的第一步。

"今天一座摩天大楼是真正的现代化的标志，因为它需要多种最新的技术设备来实现。打个比方，建造它所需的20世纪技术和工艺，就好像2500年前埃及造金字塔和2000年前中国造长城所需要的一样。"报道写道。

著名建筑师贝聿铭曾在一次采访中承认，他对建筑的热爱缘于童年时一次骑车路过静安寺路（现南京西路）。他看见工人们正在为国际饭店挖地基，以建造这幢高达22层的建筑。

继国际饭店之后，邬达克接到几个摩天楼设计项目，包括一个拟为轮船招商局在外滩建造的40层巨厦，但因为1935年银元危机和1937年淞沪抗战爆发等原因最终无一付诸实施。

而由于战乱、政治动乱和经济等原因，国际饭店建成后近半个世纪上海再未建这么高的大楼。国际饭店作为上海的"第一高度"一直持续到1983年上海宾馆建成。此后，高楼大厦像雨后春笋一样在这座昔日的摩登之都

拔地而起。"国际饭店连接了我们的过去与未来，摩天楼激励我们努力奋斗。但看到上海中心城区高楼密布，出现大面积像插花一样建高层建筑的城市景观，对于未来的城市和生活环境有什么影响？"刘刚说。

1947年时局动荡，邬达克离开上海，先赴瑞士，晚年移居美国，1958年在加州因心脏病去世。他再也没有回到这座城市，而在美国加州伯克利大学谋到教职，安静地从事一直热爱的宗教与考古研究，就这么挥一挥衣袖，仿佛上海的一切都是过眼云烟。

邬达克的名字渐渐被这座城市遗忘了，直到近年关注历史建筑的人越来越多，邬达克重新"热"了起来，甚至成为家喻户晓的"网红"人物。邬达克在上海的两个故居，番禺路别墅和延安西路达华宾馆，都设立了邬达

克纪念室。

1949年，国际饭店收归国有，如今是锦江集团旗下的一家四星级酒店。1950年，上海市测绘部门以国际饭店楼顶旗杆为原点，确立了上海城市平面坐标系。如今，酒店在大堂用一个美丽的装置将原点位置标出，供人欣赏。也许这座昔日中国第一摩天楼也是一个新的原点，让人思考上海这座城市的历史与未来。

昨天： 国际饭店　**今天：** 国际饭店　**建筑师：** 邬达克　**建筑风格：** 装饰艺术风格
参观指南： 酒店大堂可供参观，大堂夹层的走廊展示着加拿大学者提供的珍贵历史照片。

In a 1933 cartoon by famous artist Zhang Guangyu, two country bumpkins were speaking against the backdrop of the Park Hotel, then still under construction. Bumpkin A asked, "Wow! Such a tall building, what's it for?" Bumpkin B replied, "You sure know nothing, it's for when the water in the Huangpu River swells up!"

The 83.8-meter, 22-story hotel dominated the city's skyline from 1934 to 1983. It is also architect Laszlo Hudec's most famous work.

In 1930, the Joint Savings Society, founded by four Chinese banks, decided to invest in a tall modern hotel fronting the former race course. Hudec won the project after an open competition largely due to his previous work on the society's Union Building near the Bund.

"In the early 1930s, at the peak of his career, Laszlo Hudec became the protagonist of the most ambitious and important architectural adventure in the history of Shanghai--the construction of the city's first skyscraper," Italian architectural historian Luca Poncellini wrote in his book *Laszlo Hudec*.

The Park Hotel, or J.S.S. Building, was the first step toward raising Shanghai from the level of a common city to a "leading" city, reported English paper *China Press* upon the hotel's grand opening in 1934.

"A skyscraper today is the first sign of real modernity because it requires all the most up-to-date mechanical devices of man to perfect it. In other words, it is the quintessence of 20th century engineering and skill just as the Pyramids were to Egypt 2,500 years ago and the Great Wall was to China 2,000 years ago," the paper

reported.

Tongji University professor Liu Gang says Shanghai was in an atmosphere to grow higher since the 1920s under the influence of American skyscrapers and worldwide optimism before the 1929 Great Depression.

"Shanghai's booming economy, China's temporary political stability and cultural prosperity were all forces to push the city higher and higher. Some tall buildings had been erected since the late 1920s, such as the Sassoon House and Broadway Mansion. Park Hotel was one of them," he says.

During a US trip in 1929, Hudec witnessed the architectural upheaval that left leading American cities dotted with skyscrapers and grand hotels.

"He spent a long time in New York, Chicago and other cities

and made drawings of many skyscrapers and their decorative details. His trip to America must have had a decisive influence on the designs he later drew for the J.S.S. Building," says architectural historian Poncellini.

Professor Liu says New York in the 1920s had a huge visual and psychological impact on newcomers, who would naturally feel a kind of power and ambition.

The facade of the Park Hotel is emphasized with vertical stripes, which shrink layer upon layer until the top, a typical American modern Art Deco style. Today its imposing but stable silhouette, as well as the staircase-like tower above the 15th floor provide a unique elegant look compared to the surrounding modern skyscrapers.

As a visiting scholar from University of Pennsylvania in the US years ago, professor Liu says Hudec may have been inspired by the Bryant Park Hotel in New York.

"Located close to the Public Library, the Bryant Park Hotel closely resembles Shanghai Park Hotel, from the architectural features to the unique facade covered by dark-brown tiles. New York's version was a bit smaller in scale," Liu says.

"But Hudec was so brilliant that he simplified the shape of the New York hotel to suit Shanghai's context. His design highlighted architectural grandeur in a more intense way," he adds.

"He captured the essence of this architectural form and made a cultural transplantation. I don't think it was a copy. He was not obedient, instead he joined this trend with unprecedented confidence. Remember it was an era of internationalism rather than globalization."

It's widely known that Park Hotel sits on a reinforced concrete raft base of 400 33-meter-long piles of Oregon pine, which is topped by light-weight alloy with great strength to prevent it from sinkage, a problem Shanghai architects had struggled with for decades.

Compared with the efforts to solve the foundations problem, Hudec expended more energy just to get approval to build the hotel.

According to a 1934 article in *China Press*, the then Shanghai Municipal Council had not allowed buildings over the height of the J.S.S. Building. Various excuses were offered such as fire hazards or danger of sinking.

"In addition to the opposition from the government, Mr Hudec had to convince the owners of the bank that a skyscraper was not only reasonable but highly advisable. At first they were disinclined to accept the reasoning but in the end Mr Hudec was able to convince the owners that a skyscraper on Bubbling Well Road would not only be most practical but one of the most unique ventures that have been undertaken in the Far East. With the consent of the owners in his pocket, Mr Hudec returned to the fray with the Council authorities."

This took considerably more time. With promises of a fire lookout on the 22nd story and other guarantees with regard to fire sprinklers, the quality of steel and general structural materials, permission was finally given for the new skyscraper.

"It was a long and hard fight waged by architect Hudec, but eventually he won. Looking at this beautiful example of engineering skill, who today will not extend a hearty applause for the pioneering work of Mr. Hudec in fostering the first real skyscraper in the Far East. No doubt others will follow and when they do they must look back upon the Joint Savings Building as their rightful forebearer," the report said.

On the morning of Decem-

ber 1, 1934, then Shanghai Mayor Wu Te-chen cut the ribbons at the entrance and officially opened the Park Hotel. From the opening day, the hotel became a major venue of modern life and the first choice of international VIPs while in Shanghai. Renowned Chinese-American architect I.M. Pei had admitted it was the Park Hotel that greatly attracted him to architecture.

It's noteworthy that the hotel was built by main contractor Voh Kee Co. with various Chinese suppliers providing everything from black polished granite on the plinth of the external walls to dark brown Taishan tiles on the facades.

Hudec and his team enjoyed the highest honor of their career from this skyscraper. He received commissions for other skyscrapers one after another. Even the China Merchants Steam Navigation Company planned to erect a

40-story building on the Bund.

However within a year the 1935 world silver crisis had a devastating effect on Shanghai. The financial center of China fell into a depression, which was exacerbated after Japan invaded Shanghai in 1937. Construction came to an abrupt halt in the city.

After 1949, the Park Hotel became state-owned and now serves as a 4-star hotel of the Jinjiang Group.

The Park Hotel remained the city's only skyscraper for decades, until it was surpassed by the 26-story, 91.5-meter Shanghai Hotel in 1983.

Once Shanghai Hotel was built, high-rise buildings began mushrooming around Shanghai, especially in the 1990s.

The Chinese term for skyscrapers, mo tian lou, literally means magical building that reaches the sky. In his famous book *Shanghai Modern: The Flowering of a New Urban Culture in China 1930-1945*, Harvard University professor Oufan Lee describes skyscrapers as a visible sign of the rise of industrial capitalism and the most intrusive addition to the Shanghai landscape, which offered a sharp contrast to the general principles of low-rise Chinese architecture.

No wonder skyscrapers had elicited such heightened emotions about socioeconomic inequality--he high and the low, the rich and the poor--in 1930s cartoons and films such as the two country bumpkins talking about the Park Hotel.

"The Park Hotel connects our past to the future. Skyscrapers encourage us to strive for more, but now our city is filled with them like massive flower arrangements planted in a casual way," says professor Liu.

In a 1950 municipal survey, the flagpole of the Park Hotel was referred to as "Zero Center Point of Shanghai" because of its central location and height. Upon its 80th anniversary, perhaps the Park Hotel will once again become a starting point for us to ponder the city's past and its future development.

Yesterday: the Park Hotel   **Today:** the Park Hotel   **Date of construction:** 1934
Architect: Laszlo Hudec   **Architectural Style:** Art Deco
Tips: I'd suggest visit the informative archival showeroom at the second floor of the hotel.

一座金色的教堂

A "Golden Church"

慕尔堂是一座"金色的教堂"。教堂的尖券窗装有澄黄色玻璃，阴天能将黯淡光线滤成金色，在老牧师的回忆中无论晴天雨天，教堂总是"阳光明媚"。

　　这是建筑师邬达克在上海仅存的两座教堂作品之一，另一座是位于西郊的息焉堂。

　　圣约翰大学校长卜舫济（F. L. Hawks）在1928年所著的《上海简史》中，提到1843年上海开埠后很快成为新教传教的中心。"传教士们努力在这里建教堂、学校和医院。随着上海发展成为一座最大、最重要的口岸城市，它自然而然地成为了大多数基督教教派设立中国分部的地方。"

　　《邬达克》传记作者、意大利学者卢卡·彭切里尼（Luca Poncellini）提到，基督教价值观与中式传统思维逐渐融合，更多的中国人加入教会工作并进入教会高层，到1932年时全国统计有3679名中国基督教牧师和近5000名新教牧师。

　　"而邬达克就是在这段时间首次接到教堂项目的，这时距离他和父亲在匈牙利和西伯利亚战俘营设计教堂，已有15年之遥。"他在《邬达克》书中写道。

　　1929年，这位从西伯利亚战俘营逃至上海的东欧建筑师设计了拜占庭风格的息焉堂。同年，他另一件教堂作品——位于上海跑马场边上的慕尔堂开始建造。信奉路德教的邬达克还设计了位于今天希尔顿酒店位置的德国新福音教堂，可惜已被拆除。

　　慕尔堂始创于1887年，原名监理会堂，是南方监理公会在中国分部的新会堂，原址在汉口路云南路今扬子精品酒店的位置。因为美国信徒慕尔为纪念早夭的女儿捐资，教堂改名为"慕尔堂"。1917年慕尔堂成为社交会堂，由于活动增多加上教友人数增加到1200多人，旧堂渐渐不敷使用。信徒们募资在西藏路汉口路建造新堂。新址原是中西女塾，著名的宋氏三姐妹曾在此就读。

　　慕尔堂是当时中国最大的一座社交堂，除了宗教活动，还开展慈善救济和教育文化活动，设有女校和夜校。对于邬达克来说，这个复杂的任务要满足多种需求——设计一个能举办千人圣餐仪式的大

会堂，还需要小回廊和户外凉亭用于举行音乐会等活动。

彭切里尼评价，邬达克的设计介绍了一种不同寻常的表面图样，立面砌砖凹凸不平，为墙面平添了几分视觉动感。他还发现1929年6月完成的最终方案与1926年的草图相比改动很大。与传统西式教堂拉丁十字平面布局相比，慕尔堂的设计充满新意。教堂由5部分构成，中间是主会堂，四角分别是社会、教育、管理和娱乐部门，东北角还保存了中西女塾的旧楼。教堂的主入口面向上海跑马场，还设有内院。

1931年3月16日，来自近10个国家的1500名中外来宾和教友参加了教堂开幕典礼。英文《大美晚报》（*The Shanghai Evening Post*）发表评论，认为"新教堂采用学院派哥特风格，设计注重保留旧世界教堂艺术的丰富传统，但同时又费尽心思做到彻底的现代感"。

新教堂宽敞的大堂高达两层，顶部为肋骨拱顶形式，外侧有扶壁。大堂配有美国进口的木制长椅，室内墙面用黄褐色灰泥粉刷。值得一提的是，这是一座有全面防火设计的教堂，还安装了中央供暖等现代设备。

管理教堂的张刚先生透露，设计精美的教堂抗战期间被日军用作马厩，在"文革"时期又变为一所学校的体育馆，历经严重的破坏。幸运的是，损毁的圣坛根据一张历史照片修复了，照片的提供者曾是这张照片上唱诗班的一个孩子。

根据修缮教堂的上海章明建筑设计事务所提供的资料，慕尔堂由于高质量的设计和施工，虽历经损坏，但总体状态不错。修缮团队修复了外墙的深红色面砖等历史细节，还将新增的空调隐藏在墙边木箱中。

2010年春天，教堂完成大修

重新对公众开放。张刚回忆，当80岁高龄的建筑师章明在圣坛上介绍修缮工程时，许多教徒为恢复昔日美丽的教堂感到惊叹，流下高兴的泪水。

慕尔堂的钟楼高达42米，顶部装饰有一个5米高的霓虹十字架，曾经是人民广场的标志之一。章明设计了一个可旋转的水晶十字架，白天晶亮透明，夜晚在上海城市之心闪耀动人光芒。

昨天：慕尔堂　**今天：**慕尔堂　**建造时间：**1929-1931年　**建筑师：**邬达克
建筑风格：美国学院哥特式风格
参观指南：周六周日开放。建议参观邬达克另一座拜占庭风格的教堂作品——位于可乐路1号的息焉堂。

The Moore Memorial Church was a "golden church" in the memories of old priests. The pointed arch window was embellished with yellowish stained glass, which reflected vague sunlight even on a cloudy day and thus ensured a mysterious, religious atmosphere of "an always sunny church".

This is one of the two surviving local churches designed by Laszlo Hudec. F.L. Hawks Pott's 1928 book *A Short History of Shanghai* described Shanghai as a center of Protestant missionary work soon after the city opened port in 1843.

"The efforts of the missionaries were exerted in founding churches, schools and hospitals. Naturally as Shanghai developed into the largest and most important treaty port, it became the headquarters of most of the missions carrying on work in China," Pott wrote.

"It was in this period (around 1930) that Laszlo Hudec received his first commission for churches, more than 15 years after the religious buildings erected in Hungary with his father and at the Siberian prison camp," says Luca Poncellini, an architectural scholar and author of Hudec's biography.

Located at Hankou and Yunnan roads, the church was originally built in 1887 by American missionary C.F. Reid, who belonged to the Southern Methodist Church. It was renamed Moore Memorial Church after American follower J. M. Moore made a large donation to the church in 1890 in memory of his daughter.

The church came to be used as a place for social and informal gatherings in 1917 by stretching its open time. By 1925, the number of followers had increased to more than 1,200--far too many for the church to handle.

With more donations and fund-raising activities, the construction work for a new church finally kicked off at Xizang and Hankou roads in 1929 on the for-

mer site of the Mctyeire School for Girls, where the famous Soong sisters had studied.

Hudec believed in the Lutheran Church. As the chief architect of Shanghai Societas Jesus, he had designed the New German Lutheran Church (now demolished at the site of the current Hilton Shanghai hotel and Hotel Equatorial Shanghai).

As China's largest church involved in large-scale social activities, Moore Memorial Church saw charity events and education programs. The church was open daily

for children's and adult events. It also hosted a women's school and an evening school.

Poncellini indicates that Hudec faced a complicated task involving various requirements: A hall with a capacity of 3,000 people was needed for the Eucharistic ceremonies, as was a small ambulatory and an open-air pavilion for concerts and other events.

"Laszlo Hudec's design introduced here an unusual surface pattern, obtained by setting back or forth some of the bricks layered on the facades, which succeeded in adding some visual dynamism to the wall surfaces," Poncellini says. "The final plans completed in June 1929 introduced significant changes compared with the first sketches made in 1926."

The layout was innovative other than the traditional Latin-Cross form for Western churches. It is essentially composed of five parts, including a nave that could house 1,200 people in the center, and four other sections that dealt with issues relating to society, education, management and entertainment in the four corners.

An old building of the former McTyeire School was preserved at the northeast corner. The main entrance was on Xizang Road facing the former Race Course. There were two inner courts inside the church.

Around 1,500 Chinese and foreign followers from as many as 10 nationalities attended the opening ceremony in the spring of 1931. *The Shanghai Evening Post* said in its review, "The style adopted for the new church is known as Collegiate Gothic, and great care has been taken to preserve the richer traditions of the Old World cathedral architecture although equal pains have been taken to make it thoroughly modern."

The two-story hall had an open space in the eastern end with exquisitely patterned wooden seats for the choir. Enclosed on the three sides, the seats on the second floor were arranged in a layout similar to a classic Western opera house.

The Gothic vaulted ceilings finished with stone ribbings. The interior walls were in rough amber-colored stucco, and the ceiling of the auditorium was lined with acoustic-celotex sound-proofing panels. It's noteworthy that the church was entirely fireproof, had a central heating system and other modern sanitary appliances.

Manager Zhang Gang recalls that the church suffered twice severe damage — the first one was

when it was used by the Japanese army as a horse stable during China's War of Resistance Against Japanese Aggression (1937-1945) and the other was when it turned into a school's gymnasium during the "cultural revolution" (1966-1976).

The altar, which was also demolished during that period, was revived too, according to a photo donated by a local Christian, who was one of the kids in the chorus in the photo.

According to Shanghai Zhang Ming Architectural Design Firm, the church was still in good shape due to its high quality of design and construction. The restoration team just worked on how to preserve it and artfully concealed air conditioners inside elegantly designed wooden boxes along the wall. Acoustic panels and an electric organ were also added for a better sound effect.

The facade of the church showed a Collegiate Gothic style with Romanesque manners in some parts. The external walls were adorned with scarlet bricks in a textured pattern, which were repaired by hands with brick powder.

The church boasted a 42.1-meter-tall bell tower, which was one of the highest in the city at that time. The peak of the church was installed with a 5-meter-tall neon-light cross with a motor on its base, which had been a signature of People's Square.

After a restoration to repair the golden windows and cracks on the facade harmed by neighboring high-rises, the church reopened to the public in the spring of 2010.

"Amazed by the revived beautiful church, many Christians burst out in joyful tears at the opening ceremony when 80-year-old architect Zhang Ming was introducing the restoration on the altar," recalls Zhang Gang, a longtime manager of the church.

Architect Zhang Ming designed a crystal cross during the restoration, which can swirl, looks crystal-clear in daytime and glitters with scarlet lights at night, making it a glistening church in the heart of Shanghai.

Yesterday: Moore Memorial Church   **Today:** Moore Memorial Church
Address: 316 Xizang Rd M.   **Built:** From 1929 to 1931
Architectural style: Collegiate Gothic style Romanesque manners
Tips: The church is open on Saturdays and Sundays. I would suggest you visit the other surviving Hudec church, Sieh Yih Chapel, at 1 Kele Road, Shanghai's only Catholic church topped with a Byzantine dome which Hudec designed in 1929 and completed in 1931.

螺蛳壳里做电影院

Hudec's Most Difficult Job

2010年上海世博会期间，一部关于建筑师邬达克的斯洛伐克纪录片在大光明电影院举行全球首映，片名为《改变上海的男人》。这位东欧建筑师留给上海的百件建筑作品中，大光明电影院被业界认为是最有挑战的设计。

"这是我最喜欢的邬达克作品。因为基地狭长且不规则，邬达克的设计几易其稿才最终确定，真正体现了'螺蛳壳里做道场'的功力。他的建筑图纸最终为英国皇家建筑师学会所收藏。"《上海邬达克建筑地图》作者、同济大学华霞虹教授说。

这个后来被誉为"远东第一影院"的建筑其实是一个改建项目。初建于1928年的老大光明影戏院是一座古典宫廷风格的影院，因放映辱华电影《不怕死》(Welcome Danger)影响声誉而停止营业。1932年，英籍广东人卢根与美商组建联合电影公司拆除旧楼，邀请邬达克在原址设计一座新影院。

曾参与修缮大光明电影院的建筑师林沄研究历史图纸时发现，改建后的大光明大戏院用地条件极差：整个地块形状狭长而不规整，仅内部腹地稍宽，沿南京西路的街面被夹在店铺之间。在这样的地块上要设计一座豪华影剧院，观众厅还要容纳2000个座位，是相当困难的。

邬达克的解决方案非常巧妙。他在老大光明大戏院的椭圆平面基础上，将观众厅与门厅轴线做了30度扭转。两层的休息厅设计成腰果形，与流线形的门厅浑然一体，休息厅中央还点缀着灯光喷水池。走进大光明，两部气派的大楼梯直通二楼，让人完全感觉不到基地狭长的不足。

林沄还提到，当时国泰大戏院和兰心剧院等剧场都位于街道转角处，地理位置显著，而大光明大戏院的主入口在高楼林立的南京西路，三层高的建筑很难"脱颖而出"。为此，邬达克又想出妙招，在大光明南立面上增加了一个高达30.5米的玻璃灯塔，在夜晚明亮夺目，成为点睛之笔。建筑外观呈现装饰艺术风格，立面横竖线条形成对比，中间升起的高耸灯塔成为所有竖线

条的高潮，错落有致，一气呵成，为这座娱乐建筑塑造了活泼动人的形象。

1931年，邬达克的设计方案发布后，英文《大陆报》进行了详细报道。

"上海就像世界上任何一个如此规模的城市一样，为电影而疯狂。中国的其他地区都不了解好莱坞的活力，但这座大都会每年都乐于付出高昂代价，以大饱眼福地欣赏在银幕上演的滑稽动作。上海剧院的发展最近才开始跟上大众对电影的兴趣。今天，上海刚宣布建设第一个'电影圣殿'的计划，新的大光明大戏院位于静安寺路（今南京西路）。"报道写道。

1933年6月14日，更名为大光明大戏院的新影院开幕，首映好莱坞电影《热血雄心》（Hell Below），开启了一个辉煌时代。1949年前，大光明主要放映美国福克斯、米高梅等公司的原版影片，也是工部局音乐会的常驻地。这里也是第一家使用译意风(Earphone，类似同声翻译)的影院。除了放映电影，大戏院里还设有舞厅、咖啡馆和弹子房等娱乐设施。著名建筑师贝聿铭少时常去大光明电影院看电影、打撞球。在那里，他深受美国电影的影响，也被邬达克另一件代表作——国际饭店深深吸引，从此开始了做一名建筑师的梦想。

《邬达克》传记作者、意大

利学者卢卡·彭切里尼提到，邬达克设计大光明时正值他建筑师生涯的鼎盛时期。随着国际新建筑风格的出现，邬达克的设计风格也发生了重大转变，成为上海新风格建筑最引人注目的大力推动者。他的设计风格的转变最初出现在1932年建成的的真光大楼上，这座表现主义风格的办公建筑造型简洁，有哥特式尖券的造型和褐色面砖。

"具有强烈时代感的大光明大戏院于1933年6月的落成，标志着邬达克设计风格完成了彻底的转变，他的新潮设计立刻受到建筑界的广泛关注，并由此奠定了他作为上海最有影响的现代建筑师的地位。"彭切里尼写道。而一年半后高达83.8米的国际饭店落成，进一步确立了他在上海建筑史上的先锋地位。位于南京路沿线的大光明大戏院、国际饭店和1938年建成的更现代的"绿房子"（吴同文住宅），成为邬达克三件最著名的代表作。

1949年后，大光明大戏院更名为大光明电影院。2007年，林沄任职的上海章明建筑设计事务所负责修缮影院，恢复了不少历史原貌，同时升级了观影设施。

在修缮中，因为只有黑白历史照片，调查建筑细部的原始色彩是最困难的。入口大厅的天花板有三层颜色——浅黄、浅绿和金色，观众厅的天花板也有七层不同时期的颜色，从银色到各种深浅的绿色都有。

最终，修缮团队根据上海建筑史学家罗小未先生、建筑师章明和影院老员工的回忆，为大厅天花板选用"阳光氛围"的金色箔纸，为观众厅挑选了淡雅别致的浅灰绿色。

2009年1月19日，大光明电影院重新开业，金色大厅和灰绿色观众厅都受到好评。大光明再次成为一家明亮夺目的电影院。

昨天： 大光明大戏院　**今天：** 大光明电影院　**地址：** 南京西路 216 号
建造时间： 1933 年　**建筑风格：** 装饰艺术风格
参观指南： 建筑室对外开放，放映厅需购票进入。请留意 1928 年大光明大戏院的历史遗迹——位于二楼的砖砌柱，也可以欣赏大厅地面上的"邬达克密码"。

Architect Laszlo Hudec's most difficult project in Shanghai has to be the Grand Theatre.

"Hudec showed the ultimate skills in handling this project, which almost seems impossible — to design a spacious, stylish cinema on a very unusual shaped plot of land. The draft was later added to the collection of the Royal Institute of British Architects," says Tongji University professor Hua Xiahong, author of *Shanghai Hudec Architecture*.

In 1928, a cinema was built on the plot. It earned notoriety after screening the film "Welcome Danger", a movie that humiliated Chinese by portraying them as

drug dealers and robbers. The cinema was forced to close down in 1931 because the movie ignited anger among locals.

Chinese-British Lu Geng, co-founder of United Movies Co, demolished the cinema and hired Hudec to design a new one.

Lin Yun of Shanghai Zhang Ming Architectural Design Firm and chief architect of the theater's 2007 renovation project says the plot was long, narrow and irregular, which made it extremely difficult to design a luxurious 2,000-seat cinema.

He says Hudec's solution was ingenious. Graced by fancy fountains, the lobbies on the first and second floors were shaped like cashews to fit with the shape of the land plot.

Two grand staircases led people from the entrance to the second floor. The artful use of a variety of curves created a free-flowing effect inside.

"Some famous cinemas built in that era like Nanking Theatre or the Cathay Theatre were on street corners, which were prominent locations. But the Nanjing Road facade of Grand Theatre was unfortunately sandwiched between tall shops," Lin adds. "Therefore, Hudec designed a 30.5-meter-high glass lighting pillar atop the 3-story cinema, which immediately stood out from a line of shops. The cubic glass lighting pillar was particularly eye-catching at night as it contrasted with the vertical and horizontal lines on the facade--very modern."

Hudec's style changed significantly in the 1930s as new trends swept across the city. He first experimented with this change on the True Light Buildings on Yuanmingyuan Road in 1932. The Grand Theatre marked a complete transformation, which made him the most noticeable architect in Shanghai's new architectural movement.

As soon as Hudec's tentative plan was released in 1931, English newspaper *China Press* gave a detailed report:

"Shanghai is as movie-mad as any city of its size in the world.

The rest of China is oblivious to the energies of Hollywood, but this metropolis pays a huge annual toll cheerfully to feast its eyes on the antics of the silver screen. The theaters in Shanghai have improved in proportion to the interest shown in them, until today plans have just been announced for Shanghai's first 'Cathedral of the Moving Picture' — the new Grand Theatre on Bubbling Well Road."

The Grand Theatre opened its doors to the public on June 14, 1933, with the Hollywood movie "Hell Below". It immediately became on the most popular entertainment venues in towxn, going on to screen movies produced by 20th Century Fox and Metro-Goldwyn-Mayer, as well as staging concerts. It was also the first cinema to offer simultaneous translation when foreign films were screened.

The building also had a dance hall, cafe and billiard rooms. The main auditorium was shaped like a big bell and seated nearly 2,000 people on two floors--the biggest capacity of any cinema in China at the time.

Grand Theatre was renamed Grand Cinema in 1949 and continued screening movies through the 1990s (the name was changed back after the renovation). By this time, moviegoers started going to newer facilities with modern comforts and conveniences. Thus the 2007 renovation project was not just about reviving its original look, but also upgrading the facilities. Five small halls, a roof garden and a restaurant were added during the renovation project.

Architects used one of Hudec's draft plans for the cinema to help with the restoration. Archival photos also helped. Lin says getting the original colors just right was the toughest task since they had to work with black-and-white photos. His team discovered seven layers of historical fragments from the ceiling of the main auditorium--silver in the innermost layer to various shades of green.

They chose a light green tone for the ceiling after speaking with former cinema employees and Shanghai architectural historian Luo Xiaowei.

The entrance hall's ceiling had three layers of colors, buff, gentle green and gold. The restoration team finally decided to use a golden foil based on architect Zhang Ming's impression that the theater had a "golden sunny atmosphere".

"Hudec has depended largely upon a combination of lighting effects and beautiful materials to get the desired result," Lin says. "In the old days the Grand Theatre looked even more 'shiny' than today because it was designed for the city's 'modern era' — the 1930s."

The team's hard work has paid dividends. Since reopening on January 19, 2009, moviegoers have returned to the cinema in droves. It is once again one of the city's top theaters — and also happens to be the most unique.

Yesterday: The Grand Theatre   **Today:** The Grand Theatre   **Address:** 216 Nanjing Rd W.
Built: In 1933   **Architectural style:** Art Deco
Tips: Relics of the 1928 cinema have been preserved by both Hudec and architects in charge of the 2007 renovation, including exterior brick columns on the second floor (fronting No. 2 Cinema hall) and creamy white terrazzo staircase on the western side. It's interesting to find them or decode the mysterious "Hudec symbols" used extensively on the floor. Tickets are required to enter the auditorium.

宝隆医生的宝贵遗产

Dr. Erich Paulun's Heritage

　　2020的抗疫新闻几乎每天都有武汉同济医院的消息。武汉的同济医院，全称是华中科技大学同济医学院附属同济医院，打开医院官网的医院概况的页面，就是一张大胡子外国医生的照片，第一句话就是"长江之滨，黄鹤楼下，有一所海内外闻名遐迩的医院……同济医院1900年由德国医师埃里希·宝隆创建于上海。1955年迁至武汉……"

　　这位后来被上海人昵称为"大宝医生"的宝隆（Erich Paulun）1862年3月4日出生于德国，2岁时父母患肺结核双双去世，他不幸成为孤儿。由亲戚抚养长大的宝隆选择参军，1882年在基尔的皇家弗里德里希·威廉外科医学学院学习，后成为一名德国海军的上尉随舰医生。

　　1891年，他在德国海军服役期间随军舰，第一次到访上海，亲眼所见老城里卫生条件不佳，流行病和瘟疫肆虐，穷人缺医少药深受疾病之苦。宝隆深受触动，想用自己所学改变这些悲剧。他后来写信给常驻上海的德国医生策德里乌斯（Carl Zedelius），表达了自己的强烈愿望：希望用自己所有的力量和知识为中国的穷人办一家医院。这位精力充沛的德国医生是个行动主义者。为了实现自己的想法，他开始认真地做准备工作，先回国进修学习，到两家医院工作提升外科医术，并继续到大学进修，同时也为筹建医院积攒资金。

19世纪90年代初，宝隆再次来到上海，先担任策德里乌斯的助理。1899年，策德里乌斯去世后，宝隆接替了他的工作。同年，他与另一位德国医生冯沙伯（Oscar von Schab）成立了上海德医公会，起初在德国驻沪领事馆行医，随后在后来的白克路、静安寺路（今凤阳路、南京西路）买了一块地，开办了收治中国穷人的"同济医院"（Tung Chi Hospital）。宝隆任院长。同济近似上海话里"德国"的发音"deutsche"，也有

"同舟共济"的寓意。根据1909年4月3日英文《北华捷报》报道，这家成立于1900年的医院开始只有几座从德国军方购买的白铁皮房子，只有20张床位，十分简陋。同年，宝隆和策德里乌斯的女儿结婚，在上海正式安家落户。

到了1901年，医院用来自中德人士的捐款在原址建起一座红砖建筑。

一份1909年关于医院的新闻报道写道："一楼有一间药房、几个储藏室、门诊室、仪器室和手术室。主要的手术室有三张手术台，并配有消毒器、器械箱、洗手池，实际上配备了现代无菌手术所需的所有条件。手术室外面有一个设备充足的仪器室，外面是一个装有电灯浴的小房间，用于治疗风湿病人。此外，还有其他电气设备。大楼另一端的主药房与门诊室相连，德国医生每晚在这里慈善义诊50到70位病人。楼上有12间供中国付费病人使用的房间，男女病人各6间。"

同济医院对病人"区别对

待"——穷苦华人可享受免费治疗，而德国公司的中方雇员看病需要支付费用。医院得到时任德国总领事克奈佩（Wilhelm Knappe）、上海道台和一些华商的捐助，包括叶澄衷、朱葆三和虞洽卿。

在上海，宝隆实现了自己的另一个心愿：在同济医院的基础上建一所培养中国医生的同济德文医学堂。1907年10月1日，这家得到中德两国政府支持的医学堂举行了开学典礼，宝隆担任首任校长，德国领事和上海道台都出席了仪式。医学堂在白克路同济医院的对面房屋中启动，随后于1908年在今复兴中路、陕西南路购地，邀请德国建筑师Carl Baedecker设计新校区，1912年又增建了工学堂。在动荡岁月里，这所学堂历经多次起伏变迁，最终发展为一所综合性的大学，就是今天的同济大学。

不过世事难料，1909年3月5日，宝隆医生英年早逝，年仅47岁。

宝隆医生去世后，"同济医院"改名为"宝隆医院"，以示纪念。1927年，国际饭店设计师邬达克曾负责医院新楼的设计。

巧合的是，早在接手宝隆医院一年前的1926年，邬达克也在上海设计了另一家外侨捐建的医院：宏恩医院（今华东医院所在地）。在落成时，匿名出资的美国富商用两名医生的名字为宏恩

医院的两间病房命名，以表达他
"最崇高的敬意"，其中一位就
是宝隆。

　　1946年抗战胜利后，宝隆医
院更名为中美医院，1951年又
更名为同济大学附属同济医院。
1951年到1955年，同济医院分批
迁往湖北武汉，更名为"武汉医
学院附属第二医院"。1955年10
月1日，国防部颁令在汉口路515
号建立第二军医大学急症外科医
院，对外称上海急症外科医院。
1959年，上海急症外科医院与同
济医院合并，改名为第二军医大
学附属第二附属医院，使用"上
海同济医院"的名字，后更名为

上海长征医院。凤阳路上留下的
院址，成为今天长征医院所在
地。如今，长征医院在院史中也
专门介绍宝隆医院的悠久历史和
优良传统。

　　宝隆创办的同济德文医学
堂，位于复兴中路陕西南路，校
园里，德国建筑师设计的红砖古
典建筑犹存，如今是上海理工大
学的校园。校园门口的陕西南路
也一度被命名为宝隆路。

　　而在位于四平路同济大学校
园内的校史馆里，宝隆医生的雕
像就在入口处，仿佛仍在散发着
他的热力、生命力和阳光，迎接
对这段历史感兴趣的人们。2000

年，同济大学合并上海铁道大学，将铁道大学附属甘泉医院更名为同济大学附属同济医院。

迁往武汉的同济医学院和同济医院发展为华中科技大学同济医学院和其附属同济医院。新华社2020年2月20日刊登的战疫报道里提到，武汉同济医院的院训"与国家同舟，与人民共济"，也是全国医务人员驰援武汉、战胜疫魔坚定信念的写照。来自国内顶级医院的医疗队纷纷驰援武汉，同济医院成为武汉市收治新冠肺炎重症患者最多的医院之一。这些前往武汉驰援的医疗队伍中也有来自同济医院创办地上海的医疗队，他们与武汉同济医生一起真正地"同舟共济"。

昨天：宝隆医院　**今天：**上海长征医院　**地址：**凤阳路415号　**设计师：**邬达克

German doctor Erich Paulun was orphaned at the age of 2 but he had a loving heart of determinations. Today, both Shanghai Changzheng Hospital and Tongji University originated from the charitable "Tung Chee Hospital" Paulun founded in the heart of Shanghai for poor Chinese patients.

The name "Tung Chee" or "Tong Ji" represents the transcription of the word "German" or "deutsche" pronounced in Shanghai dialect. This name not only indicated the charitable hospital for Chinese was initiated by Germans, but also referred to the Chinese idiom "tong zhou gong ji" meaning "on the same boat".

Born in Pasewalk of Germany in 1862, Paulun had studied in the Friedrich Wilhelm University in Berlin, an army medical institution, and served on Germany navy ships S. M. S. Wolf and Iltis in Asia in the late 1880s and early 1890s.

In a letter to a Shanghai-based German doctor Carl Zedelius, whom he knew during navy times, Paulun shared his idea to found a charitable hospital for poor Chinese patients who were suffering from illness and without access to medicine in old Shanghai.

To realize this hospital dream, he left the navy, worked in two hospitals in Germany to improve his surgeon skills and began raising funds. In 1895 he returned to Shanghai to work as assistant to

Zedelius and became his successor after he died in 1899.

The same year Paulun and German doctor Oscar von Schab founded the Shanghai German doctors' guild and purchased a land in Burkill Road (today's Fengyang Road) across Bubbling Well Road to build the Tung Chee Hospital.

According to *the North-China Herald* on April 3, 1909, the hospital founded in 1900 "at first consisted only of a few corrugated iron buildings purchased from the German military authorities." In 1901 a brick building was erected by funds contributed by both German and Chinese residents.

"On the ground floor there are a dispensary, store-rooms, out-patients' rooms, instrument rooms and operating theatres," the 1909 news report documents.

"The main operating theatre has three tables and is equipped with sterilizers, instrument cases, washbasins and in fact with every requisite for a modern aseptic surgery.

"A well-stocked instrument-room opens out of this theatre, and beyond is a small chamber fitted with a Sanitas electric light bath for rheumatic patients, and other electrical apparatus.

"The main dispensary, at the other end of the building, connects with the out-patients' room, where the German doctors see between 50 and 70 charity patients every evening. Upstairs are 12 rooms for Chinese paying patients — six for men and the same number for women."

Professor Li says the hospital treated two kinds of patients — poor Chinese in the International Settlement were treated for free while Chinese employees working for German firms were charged fees.

The hospital was founded with the support of the then German Consul General Wilhelm Knappe who wanted to increase German influence in China through education and medical services. The hospital also received funds from Shanghai Taotai (the circuit intendant for foreign affairs in Qing Dynasty), and prominent Chinese merchants including Ye Chengzhong, Zhu Baosan and Yu Yaqing.

The hospital was renamed Paulun Hospital after the German doctor died of disease a day after his 47th birthday in 1909. In the news story regarding the change of hospital name, *the North-China Herald* says a few foreign residents, besides Germans, knew of the existence of the hospital which had become well-known to the Chinese.

"No more fitting memorial could be found to the name of one who gave up so much for others than to establish the institute for ever as the Paulun Hospital," *the North-China Herald* reported.

Unfortunately the old buildings of the Paulun Hospital do not remain today. It's possible that the huge extension part, designed in 1927 by Park Hotel architect Laszlo Hudec, is wrapped inside a modern surface added during a 1980s renovation.

When Hudec's other hospital, the Country Hospital (the No.1 building of today's Huadong Hospital), was unveiled in 1926, the

anonymous donor endowed two wards each in memory of Shanghai's departed philanthropists--the late Dr MeLeod and the late Dr Paulun "for whom the donor had the greatest respect," according to a report in *the China Press* on June 9, 1926.

In the 1950s Tongji Hospital was moved to Wuhan of Hubei Province and the site since then has been used by the Shanghai Changzheng Hospital which was attached to the Second Military Medical University.

During the last few years of his life, Dr Paulun founded Tongji German Medical School in Burkill Road hospital with support from both German and Chinese governments. The school hosted a grand opening ceremony on October 1, 1907, which was attended by representatives of German consulates and Shanghai Taotai.

Owing to tight space within the hospital, the school purchased land in today's Fuxing Road M. in 1908 and commissioned German architect Carl Baedecker to design a new campus.

In 1912 the school was expanded to include engineering in its programs and got its new name as Tongji Medical and Engineering School. In the following turbulent years, the school endured many changes and moves and eventually grew to be a comprehensive university specializing in engineering which is today's Tongji University.

In 1978, the then Tongji University president Li Guohao restored the university's relationship with Germany.

Though the Paulun Hospital buildings, along the Bubbling Well Road, have been demolished, Changzheng Hospital introduces history and heritage left by this pioneering hospital in its own hospital history today.

At the Fuxing Road campus, red-brick buildings designed by a German architect are largely preserved and used by the University of Shanghai for Science and Technology.

In the Tongji University History Museum situated in a quiet corner of its Siping Road campus, a statue of Dr. Paulun welcomes visitors at the entrance.

Yesterday: Paulun Hospital   **Today:** Shanghai Changzheng Hospital
Address: 415 Fengyang Road   **Architect:** L. E. Hudec

上海悼念宝隆医生逝世
Shanghai in mourning after philanthropic doctor dies

大家都非常遗憾地得知宝隆医生去世的消息。他几天前因感染伤寒被送医，因为肾脏并发症，他于昨天凌晨4点死于尿毒症。几年前，他创立了一家面向中国人的慈善医院，后来又创办位于白克路（今凤阳路）的中德医学院。对于上海这座城市来说，他首先是一名外科医生。众所周知，宝隆医生的昵称"大宝医生"是对他的医术和勇气的称赞。宝隆医生的敏捷和决断力挽救了很多人的生命。无论天气好坏，无论白天黑夜，他总是为了病人而随时待命，对待免费病人和有钱病人一视同仁。许多他的穷苦病人都能说出宝隆医生所做的善事，他会为病人急需的假期资助费用，对病人耐心照顾。就在他去世前不久，宝隆医生还说希望再活20年，以便继续从事他所奉献的职业。

The news of the death of Dr. E. H. Paulun will be learnt with extreme regret by the whole community. Dr. Paulun was taken to the General Hospital only a few days ago, suffering from typhoid fever. Kidney complications set in, and he died at 4 o'clock yesterday morning from uraemia.

Dr. Paulun was one of the best known Germans in Shanghai, not only to his fellow-countrymen, but throughout the entire community.

He founded a charitable hospital for Chinese, the natural corollary to which was the German Medical School for Chinese in Burkill Road. He was a Governor of the General Hospital, member of the German School committee and a committee member of the Club Concordia, in which capacity he rendered invaluable service with his suggestions regarding hygiene in the new building.

To all Shanghai, however, he was, first and foremost, a surgeon. The nickname by which he was familiarly known was a compliment alike to

his skill and nerve. Many persons owe their lives to his promptness and decision.

In good or bad weather, at any hour of the day or night, he was always at the disposal of his patients, and he treats those from whom he knew he could receive no fee with the same consideration as the wealthy. Many of his poor patients can tell of kindly acts, of money unostentatiously given them for a much needed holiday, of his care and patience during their illnesses.

Only a short time before his death Dr. Paulun said that he would have liked to live for another 20 years to carry on the profession of which he was so devoted an exponent. Though honors fell thick upon him during his career the most lasting monument of his work will be tender regard in which his memory will be held by many who had every reason to appreciate his services.

摘自 1909 年 3 月 6 日《华北捷报》
Excerpt from *the North-China Herald*, on March 6 1909

南京路的新记录

2018年6月，我应上海市黄浦区南京东路街道邀请，到上海城市规划馆参加《城市之心——南京东路街区的百年变迁》展览开幕式。没想到，参加一个活动，接到两项任务。

一个来自展览的主办单位南东街道，他们希望请我在展览期间为市民们讲一场主题为南京路的讲座。另一个邀请来自当天也出席开幕式的同济大学童明教授，他的祖父是近代著名建筑师童寯先生。童教授为祖父这一批近代中国建筑师也策划了展览，同样希望我在展览期间，到上海当代艺术博物馆讲讲上海近代建筑和建筑师的故事。

我在写南京东路专栏时曾得到南东、外滩两个街道和童教授的热心帮助，也很高兴与公众分享自己的新发现，就接受了邀请。

后来，展览和讲座都反响不错。7月28日南京路讲座后，城市规划馆的公众号撰文评价，"讲座具有时空交错的画面感"。

我探索南京路的过程确实有点"时空交错"：先从徐家汇藏书楼泛黄的英文报纸里搜寻关于南京路的原始报道，再带着问题访谈专家业主，并走进一座座南京路历史建筑，观察它们的历史细节与最新变化。

写完位于黄浦区的南京东路和人民广场后，我沿着南京路继续前行调研静安南京西路。不知不觉，从外滩到静安寺，一座楼一座楼，写完了整条"十里洋场"。风起云涌的南京路这么长，故事那么多，但一本书只能完成一个任务，所以本书先聚焦"黄浦南京路"——南京东路和人民广场。

调研过程中，我发现曾经作为上海城市象征的这条著名商业街，似乎没有一本记录它最新面貌又适宜公众的书籍。今年我开设了微信公众号"月读上海"，以"阅读建筑"为主题，分享关于上海建筑历史的动人发现，架设一座建筑保护业界与大众之间的桥梁。上海三联书店编辑杜鹃女士看到"月读上海"后来电约稿新书，我想正好尝试一下完成这个任务。

南京路给人的印象是热闹喧嚣的，读完这本书，也许可以从新的角度来发现它的美，阅读建筑故事，感受到这条历史街道深藏百年的底蕴和能量。

值得一提的是，"黄浦南京路"所在的外滩、南东街道都曾主办具有开创性的"阅读建筑"活动，我有幸参与其间。2017年5月，南东街道在全市率先为10座历史建筑安装了二维码铭牌，我在上海音乐厅主持了启动仪式暨论坛，如今建筑二维码铭牌已在全市普及。而在外滩街道的大力协调下，黄浦区在2018年6月1日推出"走进外滩.阅读建筑"项目，让公众零距离欣赏9座历史建筑之美，反响热烈，我也为此专门录制了喜马拉雅电台音频节目——《听，乔争月讲述外滩建筑》。

我2018年开始动笔写《阅读南京路》，2019年因为受邀撰写《外滩.上海梦》一书而暂时搁笔，今年春节疫情期间终于静下心来，完成书稿。在本书付梓印刷之际，我得知华为全球旗舰店在哈同大楼开业，急忙赶到现场见证并记录这南京路最新的建筑故事，加入书中。

2020年的第一天，我在"月读上海"发刊词里写到："上海是一座伟大的城市，愿'月读上海'为这座城，洒落一点温润的月光。"今年，南京路步行街将延伸至外滩，展露新的风貌。也愿这本"月读南京路"与洒满阳光的新南京路，一起呈现给喜爱这座城市的人们。

乔争月

2020年6月

武康路月亮书房

A New Record of Nanjing Road

In June 2018, at the invitation of East Nanjing Road Subdistrict Office of Huangpu District, I attended the opening ceremony of an exhibition named "The Heart of Shanghai-A Century of Changes in East Nanjing Road Block" in the Shanghai Urban Planning Exhibition Hall. Unexpectedly, I received two more invitations during the event.

The East Nanjing Road Subdistrict Office, curator of the exhibition, invited me to give a lecture on Nanjing Road during the exhibition period. Another invitation came from Professor Tong Ming of Tongji University who attended the opening ceremony that day. His grandfather was Tong Jun, a famous architect of modern China. Professor Tong also planned an exhibition for a generation of modern Chinese architects including his grandfather at the PSA Power Station of Art in Shanghai. He would like me to tell the stories of modern architecture and architects in Shanghai to audiences of that exhibition.

Since I received help from Subdistrict Offices of both East Nanjing Road and the Bund, where Nanjing Road E. was located, as well as Professor Tong, I accepted the invitations. I did have some new findings of the street to share when I was writing the column on Nanjing Road E.

Afterwards, the exhibitions and lectures were all well received. After my lecture on Nanjing Road on July 28, the official Wechat account of Shanghai Urban Planning Exhibition Hall commented, "The lecture has led us through a zigzag of time and space."

My exploration of Nanjing Road is indeed imbued with some "time and space": First I searched for the original reports about Nanjing Road in the century-old English newspapers in the Xujiahui Library, then I interviewed the experts/owners and walked into these historic buildings along the Nanjing Road, observing the historical details and the latest changes.

After finishing my series of Nanjing Road E. and the People's Square in Huangpu District, I walked forward and furthered my work on Nanjing Road W. in Jing'an District. From the Bund to Jing'an Temple, I wrote about most historical buildings along the whole "Shi Li Yang Chang" or "10-mile-long foreign metropolis". Due to limits of the book size, it is very hard to have all the stories of Nanjing Road in one book. So one task at a time, this book focuses on "Nanjing Road of Huangpu District"-Nanjing Road E. and the People's Square.

Also when I did my research for my column, I was surprised to find that this famous commercial street which is regarded as a symbol of Shanghai does not seem to have a book to record its latest developments.

This year I opened a public WeChat account named "Yue Du Shanghai" or "Moon's Reading of Shanghai" to share my amazing findings about Shanghai architec-

tural history and build a bridge between the circle of architectural conservation and the public. After reading my articles in this account, Ms. Du Juan, the editor of Shanghai Sanlian Publishing House called for the cooperation of publishing of my stories and findings, and we both thought a new record of Nanjing Road would be a nice try for the public readers. People's first impression of Nanjing Road is always hustle and bustle, but with this book, you may be able to discover its beauty from a new perspective and feel the historical meaning and vital energy of this historical street.

It is worth mentioning that both the Bund sub-district and the East Nanjing Road sub-district, where the Huangpu section of Nanjing Road is located, have hosted groundbreaking events in introducing architectural heritage to the public in which I was fortunate to take part. In May 2017, the East Nanjing Road Subdistrict Office took the lead in installing QR code nameplates for 10 historical buildings on the People's Square. I was invited to preside over the

launching ceremony and a forum at the Shanghai Concert Hall. Today, QR code nameplates on historical buildings have become popular in Shanghai. With the Bund Subdistrict Office as a major organizer, Huangpu District government launched the "Walk into the Bund and Reading the Architectures" project on June 1, 2018, which allowed the public to enjoy the beauty of 9 historical buildings on the Bund at no distance. I was invited to record an internet radio program called "Listen, Michelle Qiao is talking about the Bund buildings" at the audio platform Himalaya for this project.

I started writing this book in 2018, which was temporarily suspended in 2019 because I was invited to write the book "Shanghai Bund, Shanghai Dream". In this spring, as we all had to slow our normal work and life pace due to the outbreak of the unexpected epidemic, I took on the fight to restart and completed the work in this unusual peaceful period.

On the first day of 2020, I wrote in the launching article for Wechat account "Moon's Reading of Shanghai" that "Shanghai is a great city and I hope 'Moon's Reading of Shanghai' will sprinkle some gentle moonlight of this city." In this year, the pedestrian street of Nanjing Road will continue its extension to the Bund and showcase a new look. I hope, at that time along with the new Nanjing road splashed in sunlight, this book as a new record of Nanjing Road, can be presented to people who love this city.

Michelle Qiao
June 27, 2020
Moon Atelier on Wukang Road

参考书目

上海百年建筑史 1840-1949（第二版）伍江著，同济大学出版社，2008

上海近代建筑风格 郑时龄著，上海教育出版社，1999

上海建筑指南 罗小未主编，上海人民美术出版社，1996

上海近代建筑史稿 陈从周、章明著，上海三联书店，1988

摩登上海的象征——沙逊大厦建筑实录与研究 常青著 上海锦绣文章出版社 2011

大都市从这里开始——上海南京路外滩段研究 常青主编，同济大学出版社，2005

和平饭店保护与扩建 唐玉恩 主编 中国建筑工业出版社，2013

老上海南京路 沈寂 主编 上海人民美术出版社 2003

走在历史的记忆里——南京路 1840s-1950s 上海历史博物馆编 上海科学技术出版社 2000 年

上海城市之心 马学强 主编 上海社会科学院出版社 2017

德国建筑艺术在中国 华纳 著 Ernst & Sohn 1994

南京路 东方全球主义的诞生 李天纲 著 上海人民出版社 2009

鸿达 上海的匈牙利超现代主义建筑师 Eszter Baldavari（代表匈牙利建筑师协会）著 2019

埃德加·斯诺传 ［美］汉密尔顿 著 沈蓁等译 学苑出版社 1990

报春燕纪事 斯诺在中国的足迹 武际高 著 南海出版社 1992

上海邬达克建筑地图 华霞虹、乔争月等著 同济大学出版社 2013

邬达克 ［意］卢卡.彭切里尼、［匈］尤利娅.切伊迪 著 华霞虹、乔争月译 同济大学出版社 2013

上海外滩建筑地图 乔争月、张雪飞 著 同济大学出版社 2015

上海英租界巡捕房制度及其运作研究（1854-1863）张彬 上海人民出版社 2013

A Short History of Shanghai by F. L. Hawks Pott, D.D China Intercontinental Press，2008

The Bund by Peter Hibbard Odyssey Books & Guides，2011

My China Years: a memoir by Helen Foster Snow, 1984

China to Me a partial biography by Emily Hahn, 1944

Shanghai by Harriet Sergeant, Jonathan Cape Ltd 1990

The other side of the river: Red China Today by Edgar Snow 1962

近现代英文报刊

《字林西报》及其周末版《北华捷报》*North China Daily News & North-China Herald*

《大陆报》*the China Press*

《密勒氏评论》*Milliard's Review*

《社交上海》*Social Shanghai*

《上海泰晤士报》及其周日版 *Shanghai Times & Shanghai Sunday Times*

《大美晚报》*Shanghai Evening Post*

《以色列信使报》*Israel's Messengers*

《文汇报》《解放日报》《新民晚报》

部分图片来源
Part of the Picture Sources

Helen Foster Snow:*My China Years*:
P18

惠罗百货:
P44、P49

外滩投资发展有限公司:
P64

《以色列信使报》:
P73

华建集团历史建筑保护设计院
（摄影师：刘文毅）:
P98、P99、P100、P102

沈寂《老上海南京路》:
P120

上海时装商店:
P125、P127、P135

永安百货:
P142、P144、P149

上海历史博物馆:
P182、P183、P184

上海商贸旅游学校:
P234、P236

童明:
P247、P248、P249、P251

杨子精品酒店:
P255、P260、P262、P265

格致中学:
P286

《上海近代建筑史稿》:
P293

建筑师邢同和:
P301、P305

锦江都城青年会酒店:
P318

上海音乐厅
P326、P335

国际饭店:
P339、P347、P348、P351、P352

部分图片说明
Part of the Picture Captions

P18：
著名美国记者、《西行漫记》作者斯诺
The famous American journalist Edgar Snow, author of *Red Star Over China*

P36：
犹太富商哈同夫人罗迦陵
Luo Jialing，Hardoon's wife

P83：
上海电力公司广告
Advertisement of the Shanghai Power Company

P103：
南京东路另一个老品牌"亨达利洋行"的历史照片
An archive photo of Hope Brothers, another old brand on Nanjing Road E.

P105：
老介福的历史照片
An archive photo of Laou Kai Fook's previous shop near Nanjing Road

P120：
大陆商场的历史照片
An archive photo of the Continental Emporium from Shen Ji's book Shanghai Nanjing Road

P181：
1933 年，上海跑马总会董事长伯克尔先生以夫人的名义为总会新楼和新看台安放的奠基石。
The corner stone laid in 1933 can still be seen on the façade of the building.

P248：
美琪大戏院（手绘图）
The Majestic Theatre (hand-drawn)

P249：
中国留学生在宾夕法尼亚大学（左二：林徽因；右一：陈植）
Chinese students at the University of Pennsylvania

P251：
（上）宾夕法尼亚大学海登楼绘图教室
（下）中国建筑师学会全体会员合影
(I)The Drawing Classroom in Hayden Building at the University of Pennsylvania
(II)A group photo of all members of the Chinese Institute of Architects

P255：
设计师李蟠（保罗）先生
The designer,Mr Paul Li-pan

P265：
华人企业家 C.F.Chong 先生
C.F.Chong, presently manager of the new Yangtsze Hotel

P293：
格致公学 1928 年建成的新校舍（即现在的老大楼）
An archival photo of the story founded by a committee of both Chinese and foreigners courtesy of Gezhi High School

P301：
建筑师邢同和
Architect Xing Tonghe

P315：
李锦沛（左）在中山陵
Poy Gum Lee(Left) with a colleague fronting the grand Mausoleum of Dr. Sun Yat-sen

P339：
斯裔匈籍建筑设计师邬达克
Architect Laszlo Hudec

P369：
斯裔匈籍建筑设计师邬达克
Architect Laszlo Hudec

P371：
1928 年电影院原有的外墙砖柱一直保存至今
The original exterior brick columns of the 1928 cinema have been preserved until today (fronting No. 2 Cinema hall on the second floor).

P376：
同济德文医学堂（现上海理工大学）
File photo of Tongji German Medical School(Now Shanghai University of Science and Technology)

P378：
1908 年同济德文医学堂正门照片
File photo of Tongji German Medical School, front gate, in 1908

图书在版编目（CIP）数据

阅读南京路/乔争月著. -- 上海:上海三联书店,
2020. 7

ISBN 978 - 7 - 5426 - 7046 - 5

Ⅰ.①阅… Ⅱ.①乔… Ⅲ.①商业建筑—建筑艺术—
上海 Ⅳ.①TU247

中国版本图书馆CIP数据核字(2020)第076577号

阅读南京路

著　　者／乔争月

责任编辑／杜　鹃
封面设计／0214_Studio
版式设计／一本好书
监　　制／姚　军
责任校对／张大伟

出版发行／上海三联书店
　　　　　（200030）中国上海市漕溪北路331号A座6楼
邮购电话／021-22895540
印　　刷／上海承世印实业发展有限公司

版　　次／2020年7月第1版
印　　次／2020年7月第1次印刷
开　　本／787 X 1092　1/32
字　　数／330千字
印　　张／13.75
书　　号／ISBN 978-7-5426-7046-5／T·41
定　　价／59.00元

敬启读者，如发现本书有印装质量问题，请与印刷厂联系 021-66552038